Mathematical Modeling
and Computation of
Real-Time Problems

Mathematical Engineering, Manufacturing, and Management Sciences

Series Editor: Mangey Ram, Professor, Assistant Dean (International Affairs), Department of Mathematics, Graphic Era University, Dehradun, India

The aim of this new book series is to publish the research studies and articles that bring up the latest development and research applied to mathematics and its applications in the manufacturing and management sciences areas. Mathematical tool and techniques are the strength of engineering sciences. They form the common foundation of all novel disciplines as engineering evolves and develops. The series will include a comprehensive range of applied mathematics and its application in engineering areas such as optimization techniques, mathematical modelling and simulation, stochastic processes and systems engineering, safety-critical system performance, system safety, system security, high assurance software architecture and design, mathematical modelling in environmental safety sciences, finite element methods, differential equations, reliability engineering, etc.

Circular Economy for the Management of Operations
Edited by Anil Kumar, Jose Arturo Garza-Reyes, and Syed Abdul Rehman Khan

Partial Differential Equations: An Introduction
Nita H. Shah and Mrudul Y. Jani

Linear Transformation
Examples and Solutions
Nita H. Shah and Urmila B. Chaudhari

Matrix and Determinant
Fundamentals and Applications
Nita H. Shah and Foram A. Thakkar

Non-Linear Programming
A Basic Introduction
Nita H. Shah and Poonam Prakash Mishra

For more information about this series, please visit: https://www.routledge.com/Mathematical-Engineering-Manufacturing-and-Management-Sciences/book-series/CRCMEMMS

Mathematical Modeling and Computation of Real-Time Problems

An Interdisciplinary Approach

Edited by

Rakhee Kulshrestha, Chandra Shekhar,
Madhu Jain and Srinivas R. Chakravarthy

CRC Press
Taylor & Francis Group
Boca Raton London New York

CRC Press is an imprint of the
Taylor & Francis Group, an **informa** business

First edition published 2021
by CRC Press
6000 Broken Sound Parkway NW, Suite 300, Boca Raton, FL 33487-2742
and by CRC Press
2 Park Square, Milton Park, Abingdon, Oxon, OX14 4RN

© 2021 Taylor & Francis Group, LLC
CRC Press is an imprint of Taylor & Francis Group, LLC

Library of Congress Cataloging-in-Publication Data

Names: Kulshrestha, Rakhee, editor. | Shekhar, Chandra, 1976- editor. | Jain, Madhu, 1959- editor. | Chakravarthy, Srinivas R., 1953- editor.
Title: Mathematical modeling and computation of real-time problems : an interdisciplinary approach / edited by Rakhee Kulshrestha, Chandra Shekhar, Madhu Jain and Srinivas R. Chakravarthy.
Description: First edition. | Boca Raton : CRC Press, 2021. | Includes bibliographical references and index.
Identifiers: LCCN 2020031643 (print) | LCCN 2020031644 (ebook) | ISBN 9780367517434 (hardback) | ISBN 9780367517441 (paperback) | ISBN 9781003055037 (ebook)
Subjects: LCSH: Operations research. | Mathematical models. | Mathematical optimization. | Stochastic processes.
Classification: LCC T57.6 .M3488 2021 (print) | LCC T57.6 (ebook) | DDC 511/.8–dc23
LC record available at https://lccn.loc.gov/2020031643
LC ebook record available at https://lccn.loc.gov/2020031644

ISBN: 9780367517434 (hbk)
ISBN: 9781003055037 (ebk)

Typeset in Times LT Std
by KnowledgeWorks Global Ltd.

Contents

SECTION I   Operations Research Models

SECTION II   Soft Computing Models

SECTION III   Statistical Models

Preface

The creation of this book is a joint work of the editors and the contributing authors with expertise in the field of mathematical modeling. Mathematical modeling, which plays an important role in many walks of life, is the backbone of the mathematical and applied sciences and is a multidisciplinary one. This multidisciplinary aspect makes the mathematical modeling versatile in both theory and applications. This book is no exception to this aspect.

To create this book on mathematical modeling of the real-time system through an interdisciplinary approach, a four-person editorial committee with diverse backgrounds in mathematical sciences, applied analysis, computational mathematics, operations research, probability, statistics and simulation from different parts of the world was formed. The committee believes that mathematical modeling is both an art and science, and with fundamental methodologies and efficient computational approach, the practitioners are exposed to a wide variety of real-time systems. This book offers a content-rich platform with which readers can understand a particular model, solve problems related to the model or the methodologies employed and extend results through projects, and case studies. This book provides a collection of chapters illustrating the power and richness of mathematical modeling in delivering insight into the operation of substantial real-time systems.

A key aspect of this book has been the introduction of an interdisciplinary approach to mathematical modeling. Not only do such efforts provide tangible evidence of the utility of multidisciplinary knowledge in modeling, but the precise modeling process also attracts the active participation of interested learners.

The book is conceptually organized into three parts: OR models, soft computing models, and statistical models to a total of 15 chapters. A brief description of each chapter is as follows.

Chapter 1 presents a simulation approach to rekindle the interest in (pure) vacation models through a few comparison studies. Queueing models with vacations were introduced more than four decades ago as a way to optimize the utilization of the servers. However, to author's knowledge there is no systematic study of comparing and contrasting various vacation policies under any setup, more so, in a general context.

Chapter 2 offers an investigation to the effect of imperfect service on Markovian retrial queue by including the novel features of working vacation (WV) and balking behavior of the customers. Runge-Kutta technique is implemented to solve the governing transient equations, and various performance indices are obtained in terms of transient probabilities and validated numerically.

Chapter 3 deals with a finite capacity tandem queueing system in which nodes render the queue-dependent service to jobs. The steady-state queue size distribution of the system is computed and some important performance measures viz. expected number of jobs in the system, the throughput of the system, effective reneging rate, expected total cost, etc. are also derived for increasing the applicability of the system.

Some numerical simulations have also been performed to identify critical parameter of the studied model.

Chapter 4 analyzes the performance of a machining system having a provision of fault tolerance through redundancy and maintenance is presented by using the matrix method. Markov processes for both machine breakdown and repair are used to develop the state-dependent finite population queueing model. To depict the effects of system descriptors on the system metrics, computational results have also been provided. To ensure the efficiency of the system for desired goal, numerical results are performed in this chapter.

The key objective of Chapter 5 is to examine the queueing characteristic of the F-policy retrial queueing model with a working vacation and balking behavior of the customer. Specific queueing characteristics are defined to characterize the transient behavior of the system built, such as mean queue size, expected delay time, system cost, etc. The influence of different system parameters on particular system metrics is explored by taking numerical example.

Chapter 6 considers a discrete-time, single-server, general-service queueing system with recurrent customers where the service facility is subjected to failure while initiating the service. The system is analyzed in the early arrival set up. The generating function technique is employed to derive the required distributions. An exhaustive numerical work is done to examine the impact of the parameters on the system performance. Some of the existing results are derived as a particular case of proposed model under consideration.

Chapter 7 presents a literature survey to address the utility and potential of optimization, statistical surveys and mathematical research into post-harvest supply chains (PHSCs) to reduce losses. This study covers a systematic literature survey covering the mathematical, statistical and optimization researches performed in the field of post-harvest supply chain and critical analysis to highlight the key aspects covered in the literature and their future scope.

Chapter 8 looks at the growing global environmental regulations and their pressure on various manufacturing industries to implement proper e-waste management practices. However, from an Indian perspective, the presence of a number of barriers makes the implementation of the same a difficult task. This chapter utilizes the best-worst method (BWM) to prioritize both types of barriers. The novelty of this work lies in prioritizing the barriers based on recycler perspectives. The case of the Indian electronic sector is considered to validate the proposed methodology.

Chapter 9 investigates a $M_1^{X_1} M_1^{X_2} /G_1 G_2/1$ queue with prioritized customers along with balking behavior of the arrivals. The mean number of customers in the system, waiting time of the priority as well as nonpriority customers and other performance metrics are derived and validated numerically by taking illustration. The comparison of the numerical results obtained using analytical formulae and by fuzzy analysis, has been made. The effects of different parameters on the system performance are examined by facilitating the sensitivity analysis.

Chapter 10 focuses on finding the most sustainable energy source covering environmental, technical, economic, social, political and flexibility criteria. Energy is an essential factor for the socio-economic development of societies and nations. Energy consumption in India rose very fast in the last few decades due to industrialization and urbanization. In this chapter expert's weight is collected in linguistic terminology. A fuzzy analytical hierarchy process (AHP) approach is used to analyze the collected weights. By the analysis, solar energy is found as the most sustainable energy source in India.

Chapter 11 deals with the distributed computing system (DCS), which is currently one of the key areas of interest and discusses continuous progress of microprocessor technology and computer networks. In this chapter, a serviceable model has been evolved using K-means clustering technique to establish the system's optimum impedance time by optimal assignment of tasks based on triangular fuzzy execution time and triangular fuzzy communication time processor speed.

Chapter 12 is devoted to describe a problem-solving approach using a genetic algorithm (GA) for the 0-1 knapsack problem. The main focus of this chapter is to describe the problem solving approach using genetic algorithm (GA) for the 0-1 knapsack problem. The experiments started with some initial value of knapsack variables remain continue until getting the best value.

Chapter 13 suggests a difference-cum-exponential type estimator for estimating the population mean of the study variable by using auxiliary information when the population mean of the auxiliary variable is known. The empirical study followed by simulation study has been carried out to verify the theoretical results.

Chapter 14 proposes a methodology using the Visible Infrared Imaging Radiometer Suite - Day/Night Band (VIIRS-DNB) scan data to estimate the flood-related damage in Kerala in a rapid yet efficient manner. The results demonstrate the effectiveness of the proposed methodology for a rapid yet useful flood damage assessment.

Chapter 15 provides a brief summary on the motion of the Indian plate and its interior deformation. Along the Himalayan arc, a high-velocity gradient is observed which conforms to the rapid deformation along the plate boundary. Finally, this study argues that the past earthquakes and possible future earthquakes along the plate interior depend either upon the internal lithospheric stress or on the stress from the plate boundary (i.e., Himalayas).

Rakhee Kulshrestha
Chandra Shekhar
Madhu Jain
Srinivas R. Chakravarthy

Acknowledgements

ॐ सह नाववतु ।
सह नौ भुनक्तु ।
सह वीर्यं करवावहै ।
तेजस्वि नावधीतमस्तु मा विद्विषावहैं।
ॐ शान्तिः शान्तिः शान्तिः ॥

"May God protect us both, the teacher and the student, on our journey towards attaining knowledge. May He nourish us. May we work together with great energy. May our studies be enlightening and brilliant. May there be no hate or hostility among us. Let there be peace in me, in nature, and in the divine force."

This book was written while the editors were supported by institutes of repute namely BITS Pilani (India), IIT Roorkee (India) and Kettering University, (USA). We acknowledge sincere thanks to all academic and administrative support from these institutes. Very special thanks are offered to the Vijñāna Parishad of India, an international society dedicated to the applications of mathematics in all fields of sciences and technology.

The editors wish to thank Prof. Mangey Ram; Ms. Erin Harris, senior editorial assistant, CRC Press; and Ms. Cindy Renee Carelli, executive editor (engineering), CRC Press, for their valuable suggestions, helpful discussions and constant encouragement.

It gives us immense pleasure to extend our indebtedness to all learned reviewers to support us selflessly in reviewing all chapters with their expert suggestions and feedback. We extend our gratitude to all contributors to be part of this book with their content-rich thoughts in all chapters.

We are grateful to Prof. G. C. Sharma, Dr. R. C. S. Chandel, Prof. S. C. Malik, Prof. C. K. Jaggi and Prof. Kriti Priya, who enlightened the path of progress. We are thankful to our parents, children and especially to Mrs. Kanak Lata Jain (mother of Madhu Jain) and Mrs. Shikha Gupta (wife of Chandra Shekhar) for unfailing inspiration, encouragement and continuous cooperation. Prof. Chakravarthy acknowledges the support of his wife, Jayanthi, for her understanding during time away from home due to travel. We would like to thank Dr. Amit Kumar, Dr. Shreekant Varshney and Ms. Shruti for their many hours spent on proofreading.

At last, we are also thankful to all well-wishers who provided us the best possible support, help and back-up in completion of the work.

Rakhee Kulshrestha
Chandra Shekhar
Madhu Jain
Srinivas R. Chakravarthy

About the Editors

Rakhee Kulshrestha is faculty in the department of mathematics, Birla Institute of Technology and Science, Pilani. She has received a bachelor of science, master of science in operations research from Dr. B. R. Ambedker University, Agra. She has completed her Ph.D. from Centre for Information and Decision Sciences, B. R. Ambedkar University, Agra. There are more than 33 research publications in refereed international/national journals/proceedings and a monograph to her credit. She has participated in over 40 international/national conferences in India and abroad and visited many reputed universities/institutes in Germany, Turkey and Singapore. Her research interests include the areas of applied probability, performance analysis of communication networks, inventory and supply chain management.

Chandra Shekhar is faculty at BITS Pilani, India, and is actively involved in research in the area of queueing theory, computer and communication systems, machine repair problem, reliability and maintainability, stochastic process, evolutionary computation, statistical analysis, fuzzy set and logic. Besides attending, presenting scientific papers, and delivering invited talks in conferences and universities, he has published a number of research articles in these fields in journals of repute. He is also a member of the editorial board and reviewer of many reputed journals and doctoral research committees of many technical universities. Authorship of a textbook on differential equation is also to his credit.

Madhu Jain is presently working as a faculty member in the department of mathematics, Indian Institute of Technology Roorkee, India. She has published over 400 research papers in international journals, over 100 papers in conference proceeding/edited books, and 20 books. She was conferred with many awards in India. She was a visiting fellow of the Isaac Newton Institute of Mathematical Sciences, Cambridge, UK, during the summers of 2010, 2011 and 2014. At present, Madhu Jain is holding the post of president of Vijnana Parisad of India; secretary of Operations Research Society of India, Agra Chapter; and general secretary of Global Society of Mathematical and Allied Sciences. She has participated over 150 international/national conferences in India and abroad and visited many reputed universities/institutes in the United States, Canada, Australia, UK, Germany, France, Holland, Belgium, Taiwan and UAE. Her current research interests include performance modeling of computer communication networks (CCN), queueing theory, software and hardware reliability, supply chain management, bioinformatics, soft computing, etc.

Srinivas R. Chakravarthy, Kettering University, USA, is professor of industrial engineering and statistics in the departments of industrial and manufacturing engineering & mathematics at Kettering University, Flint, Michigan, USA. He has research experience in the areas of applied stochastic modeling, applied statistics and probability, simulation for queueing, reliability, inventory and healthcare modeling.

His research is documented in more than 125 peer-reviewed publications, and he has made more than 100 presentations (including invited and plenary) in leading national and international conferences. He has also been awarded a number of honors for his pioneering teaching and research contributions. He is an area editor for the journal *Simulation Modeling Theory, and Practice*, and he is an associate editor for the journal *IAPQR TRANSACTIONS – Indian Associateship for Productivity, Quality, and Reliability*. In addition, he serves as an advisory board member and reviewer for many professional journals and international conferences. He is the founding organizer of the International conference series on matrix-analytic methods in stochastic modeling.

Section I

Operations Research Models

1 A Comparative Study of Vacation Models Under Various Vacation Policies

A Simulation Approach

Srinivas R. Chakravarthy
Kettering University, Flint, MI, USA

CONTENTS

1.1   INTRODUCTION

In classical queueing models, the server providing services is assumed to be available at all times even when there is no customer in the system. This idle time of the server is a wasted resource for the service provider and was one of the motivating factors for the introduction of the (pure or classical) vacation models in the 1970s (see, e.g., [1,2]). The term "vacation" can be interpreted in many ways depending on the context of the application. For example, the servers (or machines) may

3

have to be rested, readjusted, or undergo preventive maintenance or be shared with other systems, among other things. Since then a number of variations in how the vacation starts and ends in the case of single- and multiple-sever systems have been introduced and studied in the literature. In the classical vacation models, the servers maybe used to provide services to systems other than the one from where they are going on vacations. Almost 30 years later, the concept of a working vacation was introduced [3] so that the vacationing servers can provide services (but at a lower rate) to the customers in the original system. Vacation/working vacation models have even more significant applications these days due to the service providers making their resources available in a dynamic way using their electronic devices. Since the introduction of the vacation and the working vacation models in queueing, several hundreds of papers have been written in the last three decades or so. It is not the purpose of this note to do a literature survey on these models, as one can refer to the survey papers [4, 5] and the book [6] to get an idea of the diversity and the applied nature of the vacation and the working vacation models. Here, due to a lack of adequate systematic study of comparing and contrasting classical vacation models under any setup, more so, in a general context, we plan to discuss a few comparisons involving the (pure) vacation models under some common vacation policies using a simulation approach. Note that we use, the adjective "pure" to point out that these models are such that the servers do not serve any customers in the current system while on vacations. Before we proceed further, we briefly summarize the types of (pure) vacation models and refer the reader to the references mentioned earlier.

1.1.1   SINGLE-SERVER SYSTEMS

a. *Exhaustive multiple adaptive vacations:* The vacation starts when the system becomes idle and the server resumes servicing only when a customer is present at the time of a vacation completion or by remaining idle (if there is no customer waiting) after completing a maximum of a finite number of vacations. Note that when this maximum is set at infinity, we get the classical vacation.
b. *Exhaustive multiple adaptive vacations with setup time:* This is similar to item (a), but with an additional requirement of a setup time before offering the first service from a vacation.
c. *Threshold models:*
   i.   *N-policy:* Starts a service from a vacation only when there are N customers waiting in the queue.
   ii. *T-policy:* Starts a service or stays idle from a vacation which lasts T units of time [This is a special case of (a) with constant single vacation].
d. Batch arrivals and apply one of the aforementioned cases.
e. *Batch services:* Batches of size in $[a, b]$ are served and when the number is less than a at the time of a service completion the server goes on a vacation. A returning server will start a service only when the batch size is least a. Any service will have at most b and at least a customers in service.

f. Finite buffer and apply any of the rules from (a) to (e).

g. *Non-exhaustive:* Here, the server may take a vacation even before the system becomes idle. The types of non-exhaustive ones considered in the literature are:

   i. *Gated:* All those who were present at the beginning of a service (offered after returning from a vacation) are served and then go on a vacation.

   ii. *Limited:* Given a service period (constant or random), the server serves until the system is idle or the service period gets over before going on a vacation.

   iii. *Decrementing:* Serve until the number left in the queue is less than a pre-specified number of those present at the beginning of service initiation.

   iv. *Bernoulli:* With a certain probability offer a service or with complement probability go on a vacation.

   v. *Pure limited (P-limited):* Take a vacation after every single service and take multiple vacations when no one is waiting at the end of a vacation.

   vi. *G-limited:* Serve only min {number in queue at a vacation completion, M} customers, where M is predetermined threshold value.

   vii. *Batch limited (B-limited):* Serve a fixed number, say, M, customers during every service period. If less than M present at the end of a vacation, go for another vacation until the queue hits at least M.

h. *E-limited:* Combines exhaustive and non-exhaustive policies as follows. Serve, say, M, customers (including any arrivals during services) or all (if less than M present or arrive during the services) before going on a vacation [$M = 1$] corresponds to P-limited and $M = \infty$ corresponds to exhaustive model]].

i. *T-limited:* Given a time duration, say, T, the server serves all those present at the beginning of a vacation or when T units of time have expired, before going on another vacation.

j. *T-exhaustive limited:* Given a time duration, say, T, the server serves until the system becomes empty or T units of time have expired, before going on another vacation.

k. *P-decrementing:* Starting from a nonempty queue (at the beginning of a service) the server goes on a vacation until the queue is one less than what it was at the beginning of a service period. Example, if there were 10 in the queue at the beginning of a service period, the server will be busy until the queue size drops to 9.

l. *G-decrementing:* Starting from a nonempty queue (at the beginning of a service) the server goes on a vacation until the queue is M less than what it was at the beginning of a service period. Example, if $M = 5$, and there were 10 in the queue at the beginning of a service period, the server will be busy until the queue size drops to 4. [Note: $M = 1$ reduces to P-decrementing and $M = \infty$ corresponds to exhaustive model].

1.1.2 Multiserver Systems

a. All-server vacation:
  i. *Synchronous vacation:* All servers go on a (common) vacation when the system becomes idle. Take multiple vacations when no one is present when (the group) returning from a vacation. In this policy, it is possible for some servers to be idle without going on a vacation.
  ii. *Asynchronous vacation:* All free servers go on a vacation independently of each other. Multiple vacations are possible.
b. *Some-server vacation:* There is a limit on how many servers can be on a vacation at any given time. This generalizes the all-server vacation policy.
c. *Threshold-based vacation*
  i. All servers go on a (common) vacation when the system becomes idle and become available for service upon returning from a vacation only if there are, say, N, customers waiting in the queue.
  ii. A subgroup of a predetermined number of idle servers can go on vacation at the moment the number of idle servers hits the predetermined number.
  iii. In addition to point (ii), the number of waiting customers in the queue should be N; otherwise, the subgroup of servers go on another vacation.

1.2   VALIDATION OF THE SIMULATION

The validation of the simulated models is very important so as to use the simulated results for more complicated models. Hence, we validated our simulated models (under various combinations of arrivals/services and vacation durations) against a few published analytical models for which the numerical results are available (see e.g., [7–9]). To our knowledge, the analytical results along with the corresponding numerical results available in the literature are for *MAP/PH/1*-type model with exponential vacations. Thus, we validated against these models. The error percentages of the analytical results and the simulated ones for the mean sojourn time in the system in the case of a single-server system with *MAP* arrivals and phase type (*PH*) services varied anywhere from 0.02 percent through 12.0 percent depending on the type of arrival and service times. For the combination of positively correlated arrivals and hyperexponential services, the error percentage is largest. However, when the simulation is run for a longer period of time, the error percentage goes down. However, the error percentages for the fraction of time the server is on vacation is very low for all scenarios.

1.3   DETAILS ON INPUT DATA FOR SIMULATION

In order to conduct our simulation study, we need to select input parameters for the arrivals, services and for the server vacation times. Although selecting the parameters for the renewal arrivals is pretty straightforward, it is not very simple when it comes to the selection of those for the *MAP* processes. In Chakravarthy [9, 10], a method was proposed to select the parameters of the *MAP* to qualitatively study the effect of correlation on system performance measures. We will use that to select the *MAP* parameters for our study here. For full details we refer the reader to [9, 10].

Let (α, T) of dimension 5 denote an Erlang distribution with rate λ_i in each of the 5 phases. That is,

$$\alpha = (1, 0, 0, 0, 0), \quad T = \begin{bmatrix} -\lambda_1 & \lambda_1 & 0 & 0 & 0 \\ 0 & -\lambda_1 & \lambda_1 & 0 & 0 \\ 0 & 0 & -\lambda_1 & \lambda_1 & 0 \\ 0 & 0 & 0 & -\lambda_1 & \lambda_1 \\ 0 & 0 & 0 & 0 & -\lambda_1 \end{bmatrix}.$$

Define

$$D_0 = \begin{pmatrix} T & 0 \\ 0 & -\lambda_2 \end{pmatrix}, D_1 = \begin{pmatrix} p_1 T^0 \alpha & (1-p_1) T^0 \\ (1-p_2) \lambda_2 \alpha & p_2 \lambda_2 \end{pmatrix}, \quad (1.1)$$

where $T^0 = -Te$ with e being a column vector of 1's. The specific values of p_1 and p_2 will be mentioned at appropriate places. For arrivals, we look at two renewals and two correlated processes. These are:

ERA: This is an Erlang of order 5 with parameter 5λ in each of five stages, so that the mean and the standard deviation of the inter-arrival times are, respectively, $\dfrac{1}{\lambda}$ and $\dfrac{0.447214}{\lambda}$.

HES: This is a hyperexponential with rates 1.9λ and 0.19λ with mixing probabilities, respectively, 0.9 and 0.1. Note that here the mean and the standard deviation of the interarrival times are, respectively, $\dfrac{1}{\lambda}$ and $\dfrac{2.24472}{\lambda}$.

NCA: This is *MAP* with parameter matrices (D_0, D_1) given in Eq. (1.1) by taking $p_1 = p_2 = 0.01$, $\lambda_1 = 2.75\lambda$, and $\lambda_2 = 5.5\lambda$. Verify that here the mean and standard deviation of the inter-arrival times are, respectively, $\dfrac{1}{\lambda}$ and $\dfrac{1.0082}{\lambda}$. Further, the correlation coefficient between two successive inter-arrival times can be verified to be -0.6454.

PCA: This is *MAP* with parameter matrices (D_0, D_1) given in Eq. (1.1) by taking $p_1 = p_2 = 0.99$, $\lambda_1 = 2.75\lambda$, and $\lambda_2 = 5.5\lambda$. Verify that here the mean and standard deviation of the inter-arrival times are, respectively, $\dfrac{1}{\lambda}$ and $\dfrac{1.0082}{\lambda}$. Further, the correlation coefficient between two successive inter-arrival times can be verified to be 0.6454. The value of $\dfrac{1}{\lambda}$ will be normalized to a value required in the examples discussed later in this chapter. This normalization is needed to compare various arrival processes on a common ground. However, observe that these processes are qualitatively different in that they have different standard deviations and correlation structure.

For service times, we look at the following special cases of a *PH*-distribution.

ERS: This is an Erlang of order 5 with parameter 5μ in each of five stages, so that the mean and the standard deviation of the service times are, respectively, $\dfrac{1}{\mu}$ and $\dfrac{0.447214}{\mu}$.

HES: This is a hyperexponential with rates 1.9μ and 0.19μ with mixing probabilities, respectively, 0.9 and 0.1. Note that here the mean and the standard deviation of the service times are, respectively, $\dfrac{1}{\mu}$ and $\dfrac{2.24472}{\mu}$.

For vacation times, we look at the following two special cases of a *PH*-distribution.

ERV: This is an Erlang of order 5 with parameter 5γ in each of five stages, so that the mean and the standard deviation of the service times are, respectively, $\dfrac{1}{\gamma}$ and $\dfrac{0.447214}{\gamma}$.

HEV: This is a hyperexponential with rates 1.9γ and 0.19γ with mixing probabilities, respectively, 0.9 and 0.1. Note that here the mean and the standard deviation of the service times are, respectively, $\dfrac{1}{\gamma}$ and $\dfrac{2.24472}{\gamma}$.

1.4 COMPARISON OF GATED, *G*-LIMITED, AND *T*-LIMITED POLICIES AGAINST THE CLASSICAL VACATION (SINGLE-SERVER SYSTEM)

The purpose of this section is to point out the effect of the mean waiting time in the system by looking at three vacation policies, namely, gated, *G*-limited and *T*-limited against the classical vacation policy. For our illustrative example, we have fixed $M = 50$ and $T = 240$ units of time. The types of arrivals, services, and vacations used are indicated using the notation set forth in section 1.3. We fix $\lambda = 1$, $\gamma = 1$, and take $\mu = \dfrac{1}{\rho}$.

In Figure 1.1, for each of the three policies, we display the ratio of the mean sojourn time in the system over the corresponding classical vacation model under different scenarios. A quick note on the identifiers in the figure. *EEE* denotes the arrivals occur according to *ERA* process, the service times are *ERS* and the vacation times are *ERV*. Similarly, *NHH* corresponds to negatively correlated arrivals, namely, *NCA*, hyperexponential services (*HES*) and hyperexponential vacations (*HEV*). It should be clear with the other notation in Figures 1.1 and 1.2.

A few key observations on the ratio of the mean sojourn time in the system from this figure are as follows: (a) For the particular choice of the values of M and T, we notice that for all scenarios, the *T*-limited policy appears to be not that significant when going from a moderate traffic load to somewhat heavy one; (b) this ratio appears to be significantly large for $\rho = 0.95$ as compared to $\rho = 0.80$ for *G*-limited vacation policy under all scenarios; (c) for *ERA* and *NCA* arrivals with *ERS* services (irrespective of the type of vacation), the gated vacation policy appears to produce a higher ratio for $\rho = 0.95$, while for all other scenarios it is the *G*-limited policy that has the higher ratio.

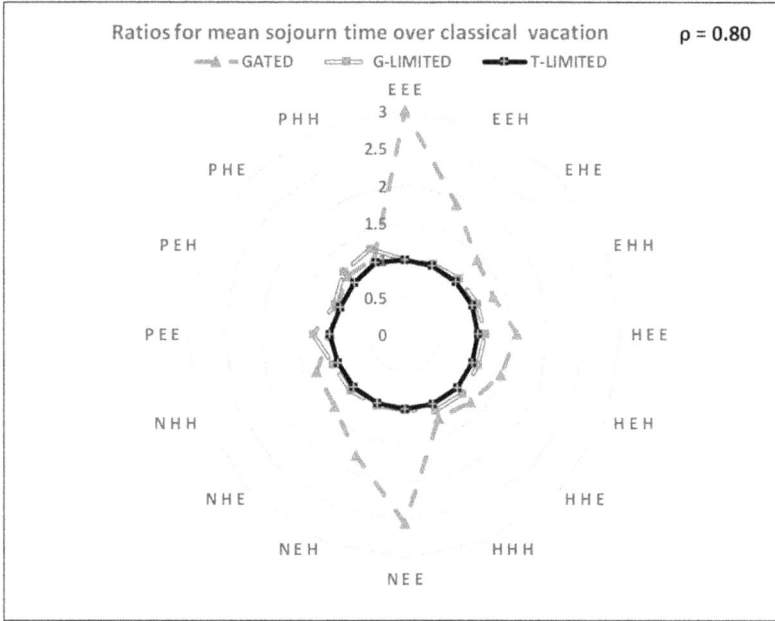

FIGURE 1.1 Ratios of the mean sojourn times of various vacation policies over the classical vacation when $\rho = 0.80$.

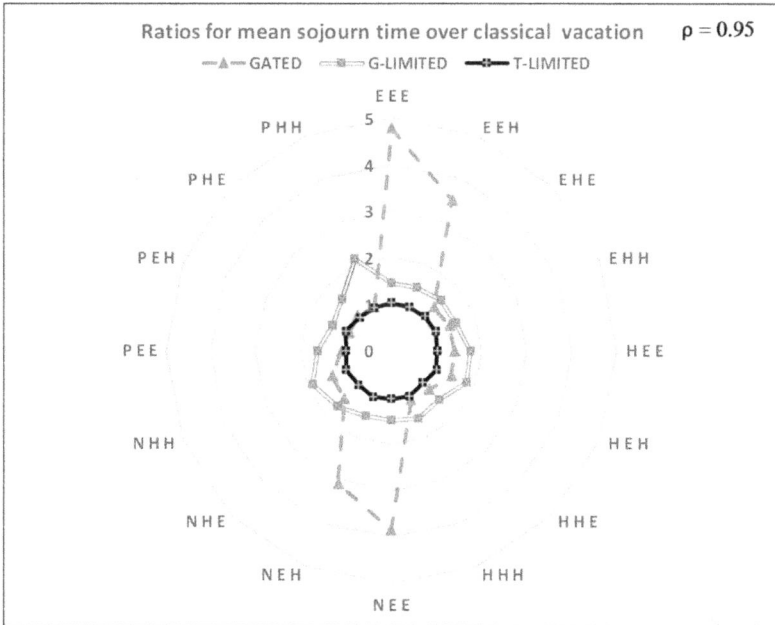

FIGURE 1.2 Ratios of the mean sojourn times of various vacation policies over the classical vacation when $\rho = 0.95$.

1.5 INTEGRATION OF CLASSICAL, GATED, *G*-LIMITED, AND *T*-LIMITED POLICIES AND COMPARING THEIR RESPECTIVE TIMES ON VACATION (SINGLE-SERVER SYSTEM)

In section 1.4, we looked at a comparison of the vacation models under various policies in an independent manner. That is, each vacation policy is simulated on its own and compared by making sure the parameters of the underlying distributions are kept the same. However, in this section, we will integrate the vacation policies seen in that section into one vacation model. Thus, there is some dependence in that the server may take different types of vacation at different points in time. That is, the server may go on a vacation by emptying the system at certain point in time, at some other time the server may go on a vacation after having completed a fixed number of services, at other times the server may possibly go on a vacation after serving continuously for more than T units of time, and so on. Thus, the purpose of this section is to look at how various vacation policies under one vacation model behave by looking at the fraction of time spent vacationing. Again, to our knowledge, this type of integration is not studied in the literature thus far.

We now consider two types of services, namely, *ERS* and *HES*, and fix the vacation times to be *ERV*. We use the same set of parameter values for the arrivals, services, and vacations. The value of M is varied from 10 to 25, and T is varied in proportion to M. That is, $T = aM$, with $a = 1, 1.5, 2$. This is mainly to contain the number of scenarios to be compared. In Figures 1.3–1.6, we display the fractions of times the server is on various types of vacation, under different scenarios. Some key observations are registered in the following text.

- For the case when $\rho = 0.50$,
  - For *ERA* arrivals, the gated vacation is the one that is used the most when services are Erlang (i.e., *ERS*); however, for hyperexponential services, it is the classical vacation that appears to be the most used by the server. This is the case irrespective of the values of M and T considered here.
  - For *HEA* arrivals, we see that the classical vacation is the one that is used the most irrespective the type of services and the values of M and T; the next most used type of vacation is the gated one.
  - For *NCA* arrivals, we see that the classical vacation is the one that is used in all scenarios. While the gated vacation is also taken in equal proportion for most of the cases when the services are *ERS*, the classical vacation is the most used when services are *HES* irrespective of the values of M and T.
  - For *PCA* arrivals, it is the classical vacation that is being used the most irrespective of the type of services and the values of M and T. Further, comparing with all other arrival processes, we notice that the positively correlated arrivals appear to be adapting the classical vacation the most.

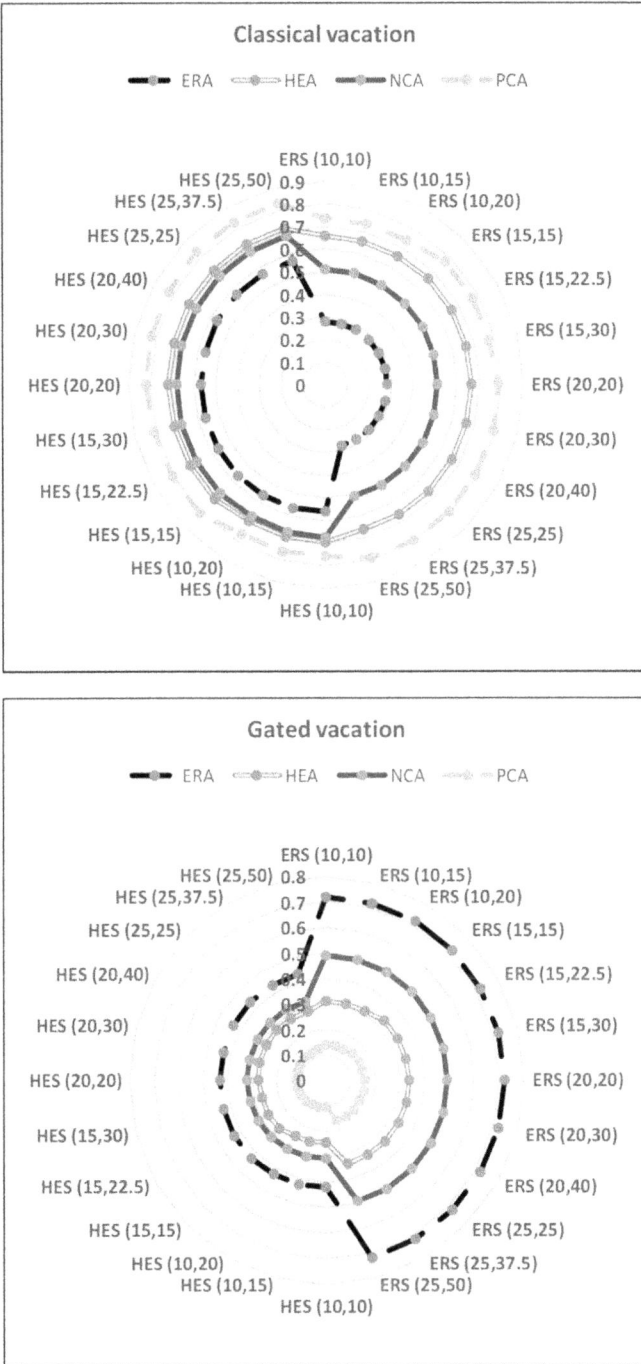

FIGURE 1.3   Fraction of time spent in vacation under classical and gated vacations for selected scenarios by fixing *ERV* for vacation duration when ρ = 0.5.

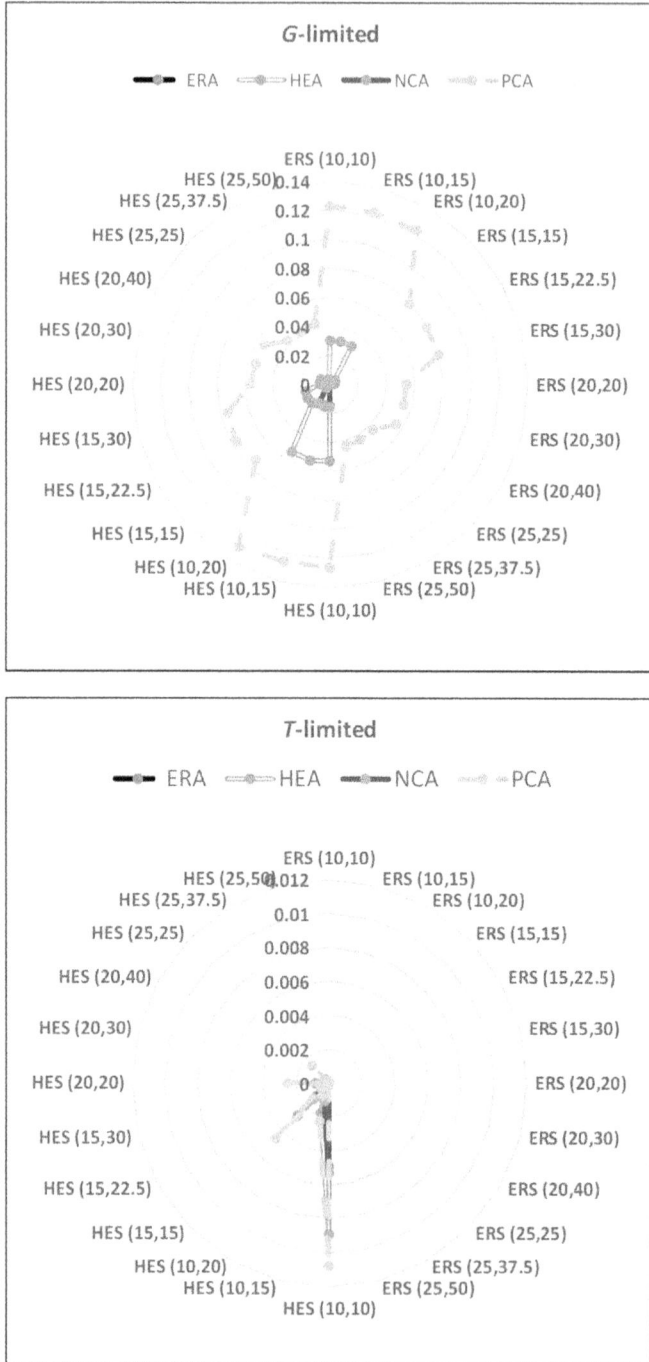

FIGURE 1.4 Fraction of time spent in vacation under G-limited and T-limited vacations for selected scenarios by fixing *ERV* for vacation duration when $\rho = 0.5$.

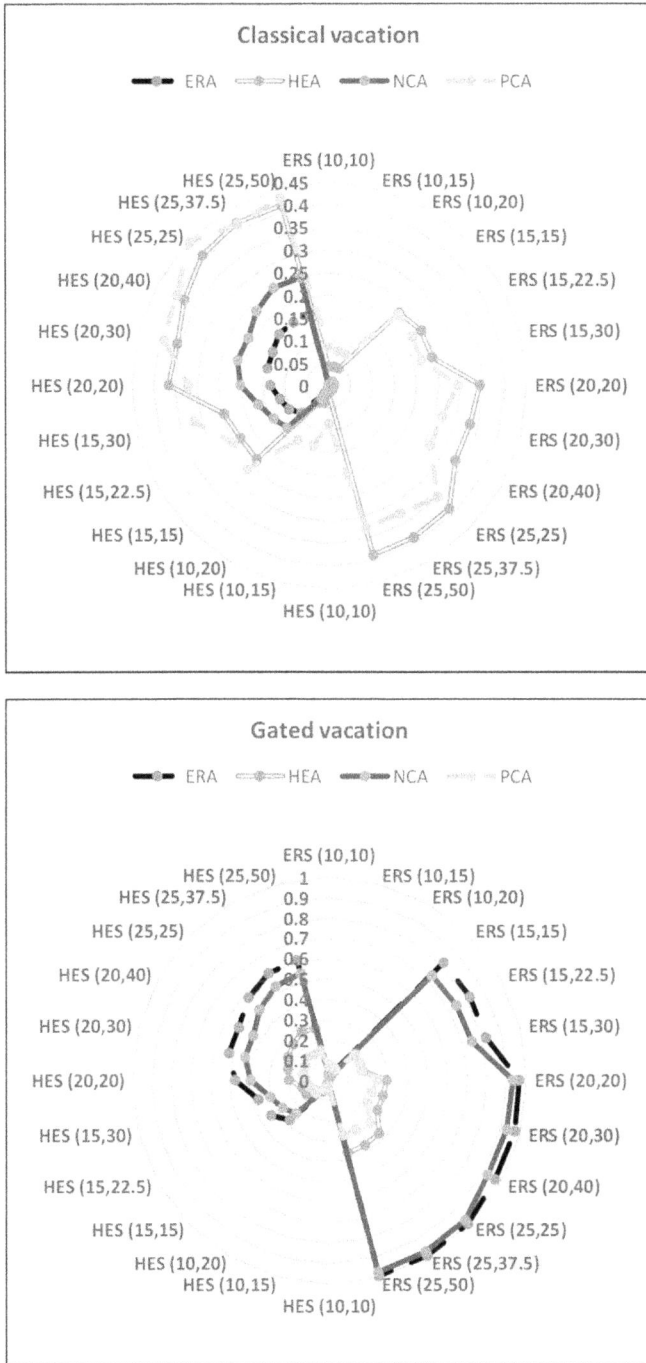

FIGURE 1.5 Fraction of time spent in vacation under classical and gated vacations for selected scenarios by fixing *ERV* for vacation duration when ρ = 0.9.

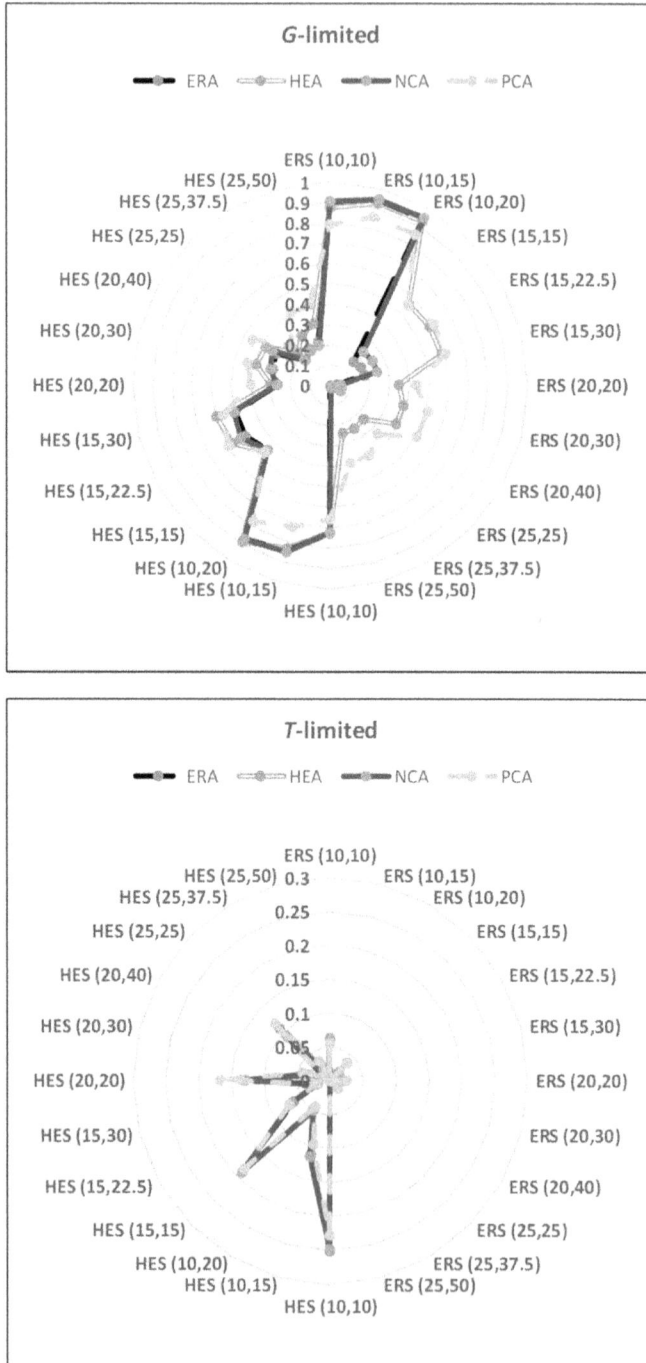

FIGURE 1.6 Fraction of time spent in vacation under *G*-limited and *T*-limited vacations for selected scenarios by fixing *ERV* for vacation duration when $\rho = 0.9$.

- For the case when $\rho = 0.90$,
  - For *ERA* and *NCA* arrivals, the G-limited vacation and the gated vacation compete with each other in the sense that for some scenarios the G-limited policy is the one used the most and in other scenarios it is the gated vacation. Thus, these two vacations appear to be the ones used the most.
  - For *HEA* and *PCA* arrivals, we notice that the G-limited policy being used the most under many scenarios for which $M \leq 20$ and for other scenarios when $M > 20$, we see that the classical vacation policy is being adopted.

1.6 COMPARISON IN THE CONTEXT OF MULTISERVER SYSTEMS

In this section, we look at the classical vacation model in the context of a multiserver system. The model considered here as follows. The system has c servers and offer services on a first-come-first-served basis. Whenever the system becomes idle, a group, say, d, $1 \leq d < c$, servers takes a synchronous vacation for a random amount of time. Thus, the system will have at least $(c - d)$ servers at any given time. The vacationing servers will be available for servicing only when at least one customer is waiting in the queue (i.e., the system has at least $c - d + 1$ customers in the system with all $(c - d)$ servers busy and the rest waiting in the queue). Otherwise, the group of d servers will keep taking vacations. The purpose of this section is to see the effect various vacation policies under one vacation model involving a multi-server system. Towards this end we will focus on the fraction of time spent vacationing. We use the same set of parameter values as in section 1.5, except for the service rate, which is taken as $\mu = \dfrac{1}{c\rho}$ and the number of servers, c, is taken as 5.

In Figure 1.7 we display the fractions of vacation under some selected combinations of arrival, service, and vacation distributions. We look at five values for d and

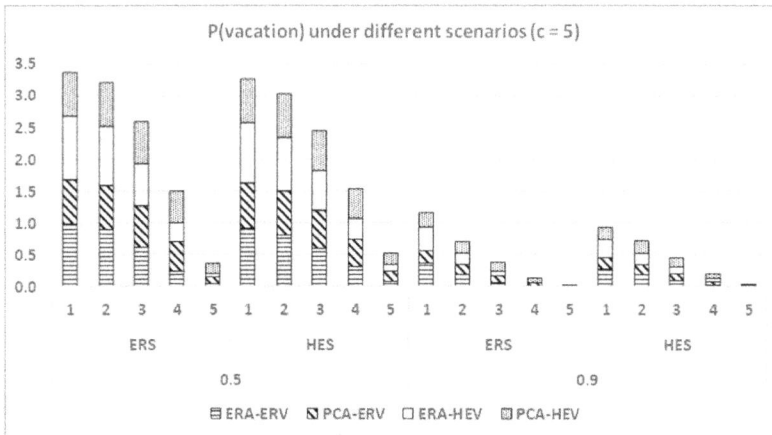

FIGURE 1.7 Fraction of time spent under vacation under various scenarios.

two values for ρ by taking $d = 1, 2, 3, 4, 5$, and $\rho = 0.5, 0.9$. A few key observations (some based on additional graphs which are not displayed here) are as follows.

- As is to be expected, increasing d decreases this probability. This is intuitively clear since the probability of waiting for d servers to become free to take a vacation together becomes smaller as d increases. This is the case for both *ERA* and *PCA*. However, the rate of decrease is higher for *ERA* as compared to *PCA*.
- The higher the variability in the services, the smaller the probability of vacation is.
- We see an interesting pattern in this probability when we compare *ERS* against *HES*. As d increases, there is a cut-off point, say, d^*, such that for $d \le d^*$, *ERS* has a higher probability and for $d > d^*$, *HES* has a higher probability.

1.7 COMPARISON OF INTEGRATED VACATION MODELS UNDER THE ASSUMPTION OF SETUP DELAY (SINGLE-SERVER SYSTEM)

So far, we looked at vacation models both in the context single and multi-server systems where there is no setup time required for the server(s) to start serving again after a vacation. In this section we will focus on having a setup time for the server returning from vacation. We use the following two distributions for the setup time.

ERU: This is an Erlang of order 5 with parameter 5η in each of five stages, so that the mean and the standard deviation of the inter-arrival times are, respectively, $\dfrac{1}{\eta}$ and $\dfrac{0.447214}{\eta}$.

HEU: This is a hyperexponential with rates 1.9η and 0.19η with mixing probabilities, respectively, 0.9 and 0.1. Note that here the mean and the standard deviation of the inter-arrival times are, respectively, $\dfrac{1}{\eta}$ and $\dfrac{2.24472}{\eta}$.

Using the same set of parameter values as seen in section 1.4 and fixing the vacations to be *ERV*, we display in a number graphs (see Figures 1.8–1.11) which cover a variety of scenarios. Note that we display representative ones to avoid duplication of similar graphs.

In the case when $\rho = 0.50$, while *ERA* and *NCA* have the largest fraction for the gated vacation policy, it is the classical vacation policy that has the largest vacation for *HEA* and *PCA*. These are for *ERV* vacations and irrespective of the service times.

In the case when $\rho = 0.90$, we see that for both *ERA* and *NCA* arrivals, depending on the type of services and the values of M and T, the fraction under the G-limited policy or the fraction under the gated policy is the largest; however, for *HEA* and *PCA* arrivals, the G-limited policy appears to have the largest fraction irrespective of the type of services and the values of M and T. The values of the fraction, of course, depend on the type of services and the values of M and T. These observations are for *ERV* vacations.

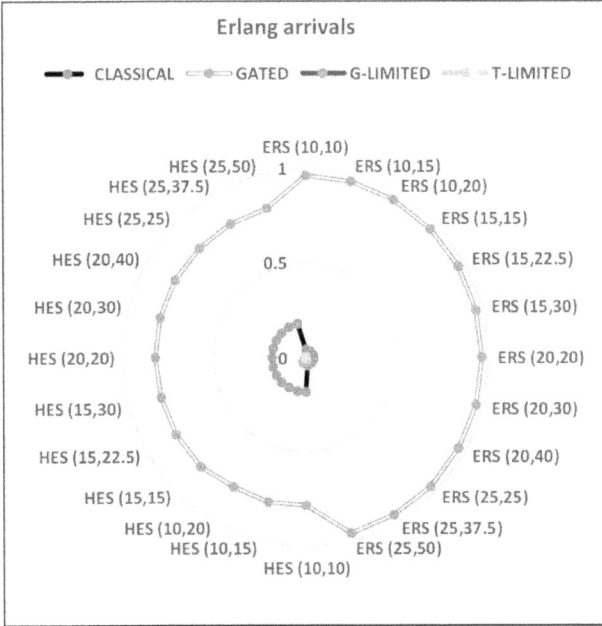

FIGURE 1.8 Fraction of time spent under vacation under various scenarios for *ERA* and $\rho = 0.5$ – Erlang setup times.

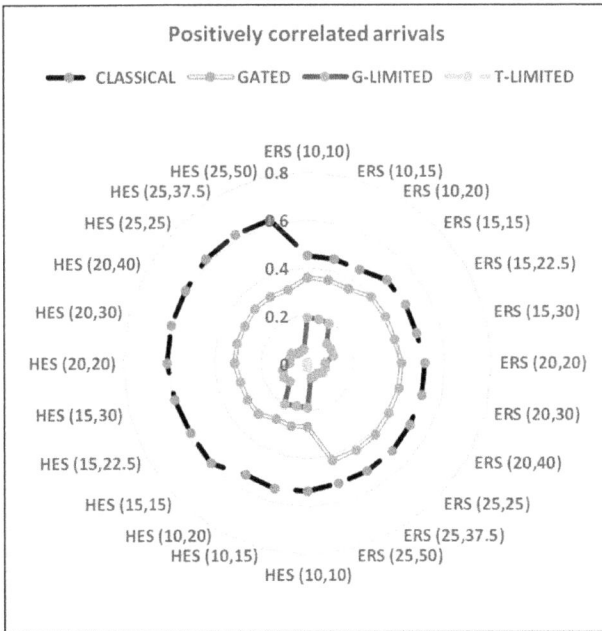

FIGURE 1.9 Fraction of time spent under vacation under various scenarios for *PCA* and $\rho = 0.5$ – Erlang setup times.

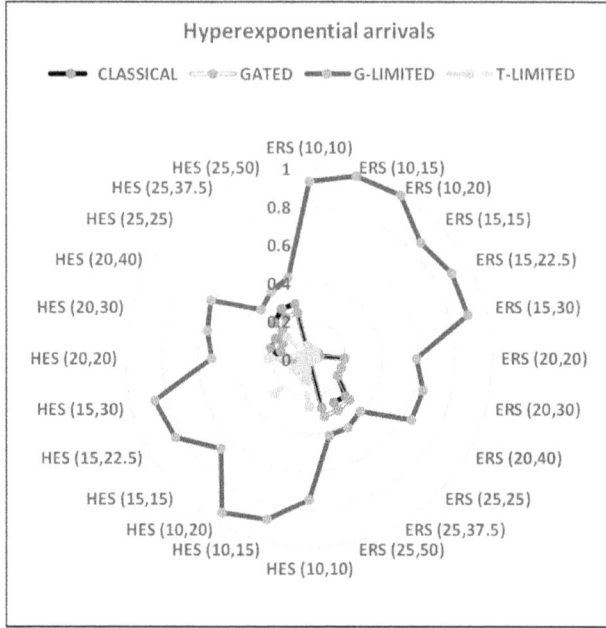

FIGURE 1.10 Fraction of time spent under vacation under various scenarios for *HEA* and ρ = 0.9 – Erlang setup times.

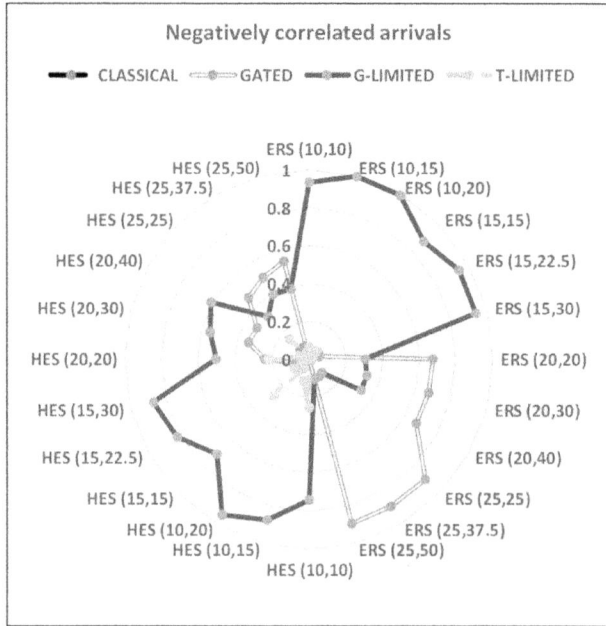

FIGURE 1.11 Fraction of time spent under vacation under various scenarios for *NCA* and ρ = 0.9 – Erlang setup times.

The aforementioned observations made under Erlang setup times also hold good for hyperexponential setup times; however, the values of the fractions differ with hyperexponential setup times producing the smaller ones.

1.8 COMPARISON OF THE CLASSICAL VACATION MODELS UNDER THE ASSUMPTION OF SETUP DELAY (MULTISERVER SYSTEM)

In this section, we look at a multi-server vacation model in the context of a setup delay. We use the same set of parameters and with $c = 5$ and investigate the behaviour of the fraction of time spent on classical vacation. Note that like in section 1.6, a group of d servers go on a vacation and resume service when at the time of returning from vacation at least one customer is waiting in the queue.

In Figures 1.12 and 1.13, we display the fraction of time spent under classical vacation for selected scenarios. From our experience (as well as the others not displayed here), we observed that *PCA* arrivals have the least fraction (compared to the others) when the group of servers going on vacation is small and becomes larger when the group size becomes larger. This appears to be the case irrespective of the type of setup times considered.

1.9 CONCLUDING REMARKS

In this chapter we looked at the classical vacation models under various vacation policies both in single and multi-server systems. To the best of our knowledge, there are no analytical/simulated models considering the integration of many vacation policies. Through simulation we analyzed the models under various scenarios and reported some interesting observation. A more detailed study is warranted to look into some additional measures. Further, the simulated studies need to be carried out under other vacation policies not covered in this chapter.

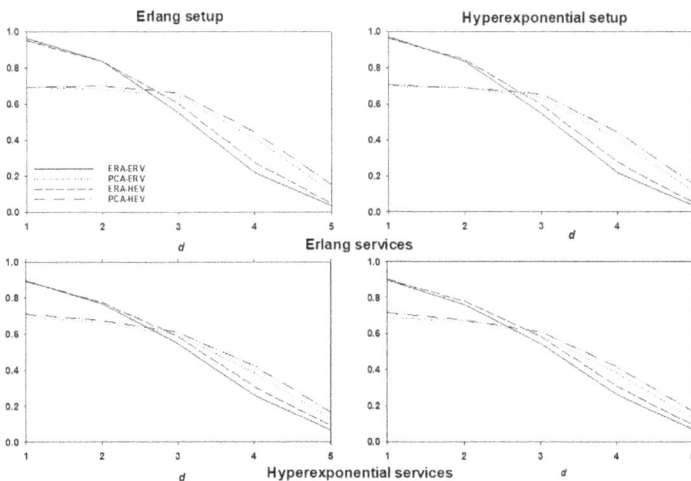

FIGURE 1.12   Fraction of time spent under vacation for various scenarios when $\rho = 0.5$.

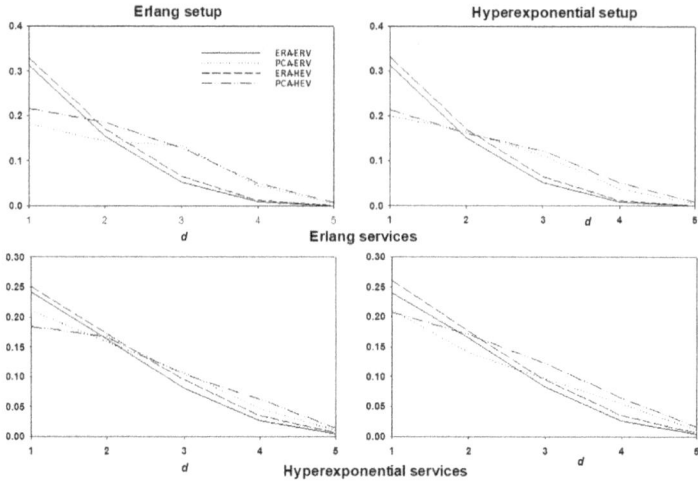

FIGURE 1.13　Fraction of time spent under vacation for various scenarios when $\rho = 0.9$.

REFERENCES

1. Levy, Y. and Yechiali, U. 1975. Utilization of idle time in an *M/G/1* queueing system. *Management Science*. 22: 202–211.
2. Levy, Y. and Yechiali, U. 1976. An *M/M/s* queue with servers' vacations. *INFOR: Information Systems and Operational Research*. 14(2): 153–163, DOI: 10.1080/03155986.1976.11731635.
3. Servi, L. and Finn, S. 2002. *M/M/1* queue with working vacations (*M/M/1/WV*). *Performance Evaluation*. 50: 41–52.
4. Chandrasekaran, V., Indhira, K., Saravanarajan, M. and Rajadurai, P. 2016. A survey on working vacation queueing models. *International Journal of Pure and Applied Mathematics*. 106: 33–41.
5. Doshi, B. T. 1986. Queueing systems with vacations – a survey. *Queueing Systems*. 1: 29–66.
6. Tian, N. S and Zhang, Z. G. 2006. *Vacation Queueing Models: Theory and Applications*, Springer, New York, NY.
7. Chakravarthy, S. R. 2013. Analysis of *MAP/PH_1,PH_2/1* queue with vacations and optional secondary services. *Applied Mathematical Modeling*. 37: 8886–8902.
8. Chakravarthy, S. R., Goel, S. and Kulshrestha, R. 2020. A queueing model with server breakdowns, repairs, vacations, and backup server. *Operations Research Perspectives*. 7: 1–13 (100131).
9. Chakravarthy, S. R. 2020. Queueing models in services – analytical and simulation approach. To appear in advanced trends in queueing theory. In *Mathematics and Statistics book series*, ed. V. Anisimov and N. Limnios: Sciences, ISTE & J. Wiley, London, UK.
10. Chakravarthy, S. R. 2010. Markovian arrival processes. Wiley Encyclopedia of Operations Research and Management Science. Published Online: 15 June 2010. https://onlinelibrary.wiley.com/doi/abs/10.1002/9780470400531.eorms0499.

2 Transient Analysis of M/M/1 Retrial Queue with Balking, Imperfect Service and Working Vacation

Madhu Jain, Sibasish Dhibar
Indian Institute of Technology Roorkee, Uttarakhand, India

CONTENTS

2.1 INTRODUCTION

In routine life as well as commercial/industrial scenarios, it has been observed that the queues of customers/jobs are built up. There are numerous examples of waiting in the queues, in particular when the server is most often busy. Sometimes the arrivals would like to wait in the retrial orbit from where they can make reattempts for getting the service after an arbitrary interval of time. When the server becomes free, then the retrial customer can get the service. If they are not satisfied, then they may demand additional service. The retrial model can be fitted in the queueing situations where customers are allowed to wait in the retrial orbit; such problems are commonly seen at ATMs, hospitals, call centres, fuel stations, restaurants, bank and railway counters, admissions in educational institutes, etc. The retrial queues with a vacationing server have many applications, including in telephone network systems, internet, distributed communication systems, manufacturing systems, etc. [1–3].

The server's vacation feature has been embedded in many queueing models to investigate the real-time system having vacationing servers. The transient analysis of a Markovian queue with a single server with discouraged customers by conserving the multiple vacation (MV) policy was investigated by Ammar et al. [4]. Servi and Finn [5] first gave the concept of working vacation (WV) to analyze M/M/1 model by including the concept that a server can work despite remaining idle during the vacation. Over the years, many researchers have done a lot of work using the WV concept in the queueing model dealing with different congestion problems. Ezeagu et al. [6] proposed the transient analysis to obtain some performance indices of Markovian single-server queue by incorporating the concept of WV and recovery policy. In queueing literature, very few research articles presented the study on the transient behaviour of Markovian queueing models by including the WV concept [7–9].

Balking concept in the queueing model is also an important characteristic from the customers' view point, and it has an adverse effect on the grade of service of any service system. If the customer notices many other customers already in a queue, then they may have the choice either to enter in the queue or to leave the system. If the customer leaves the system without joining the queue, then it is called balking. Markovian queueing model using the balking behaviour of the customers in transient setup has also been studied in a few research articles [10–12]. The transient analysis with the provision of single vacation in a feedback Markovian queue under interrupted closedown was studied by Azhagappan and Deepa [13]. By including the optimal N-policy, Azhagappan and Deepa [14] extended their previous work [13] on the transient Markovian single vacation queue with feedback. Recently, Jain and Rani [15] contributed toward unreliable server retrial queue in Markov setup having the balking and reneging behaviour of the customers.

It is observed in many service systems that the server can work at a slow speed instead of remaining idle, even when availing the WV. The sever switches to working mode when there are no customers, and as such it remains free until some customers join the system during WV. During normal busy mode, the server becomes free as soon as it completes the service and the customers from retrial orbit make reattempts for service. However, the customers may not be satisfied with the slow service rendered by the server during WV, and as such they may demand for additional service. Motivated by this fact, in this chapter we are concerned with the transient study of Markovian queueing model in general setup by including the concepts of WV, imperfect service during WV and balking behaviour of the customers. The remaining findings of the research works are arranged section-wise as follow. In section 2.2, we provide the description of the concerned model by making requisite assumptions. After that in section 2.3, Kolmogorov-Chapman equations for different system states are formulated. In section 2.4, we obtain several queueing indices in terms of transient probabilities. In section 2.5, the cost function is framed. In section 2.6, we discuss the sensitivity analysis after computing the numerical results. Finally, conclusions and future directions of research have been given in section 2.7.

2.2   MODEL FORMULATION

In the present study, we consider the WV and imperfect service concepts to develop finite single-server Markovian model for the retrial queue along with the concept of balking behaviour of the customers. The customers arrive in the system in Poisson

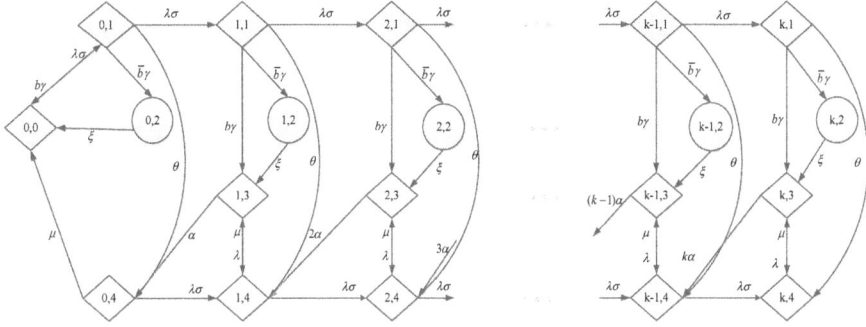

FIGURE 2.1   State transition diagram.

fashion with rate λ. On arrival, the customers join the queueing system with probability σ. In the regular busy state of the server, the customers are served following
exponential distribution (Exp-D) with mean $1/\mu$. When the system becomes empty,
the WV of the server starts; the duration of WV is governed by Exp-D with mean
$1/\theta$. In the WV duration, the server can also render service to the customers according to Exp-D with a slower rate $\gamma(<\mu)$. During the WV period, the customers are
satisfied by the primary service with probability b, and unsatisfied customers opt for
the additional service with probability $\bar{b}(=1-b)$. After getting the additional service, which is Exp-D with rate ξ, the customer leaves the system forever. During the
regular busy period, if the new arrival observes that the server is occupied, then he
or she joins the retrial orbit. The customers from the retrial pool can retry following
Exp-D with rate $n\alpha$, where n is the orbit size.

At time t, $R(t)$ and $\varsigma(t)$ denote the number of customers present in the queue
and status of the server, respectively. The state $\varsigma(t)=0$ refers that server is in WV
mode and the server is free, the state $\varsigma(t)=1$ refers that server is in WV mode and
rendering primary service to the customer, $\varsigma(t)=2$ refers that the server is providing additional service to the unsatisfied customers during working vacation, $\varsigma(t)=3$
refers that the server is free while operating in regular busy mode, $\varsigma(t)=4$ refers
that the server is in regular busy mode and rendering service to the customer. Now,
$\{(R(t),\varsigma(t)),\ t>0\}$ represents a two-dimensional (2D) bivariate stochastic process.
Figure 2.1 depicts the incoming and outgoing transition rates of all states of the
proposed model.

2.3   GOVERNING EQUATIONS

Using the appropriate rates as shown in Figure 2.1, the governing equations for different states are framed as follows:

i. The server being in WV mode but in free state $\big(\varsigma(t)=0\big)$.

$$\frac{dp_{0,0}(t)}{dt} = -\lambda\sigma p_{0,0}(t) + b\gamma p_{0,1}(t) + \xi p_{0,2}(t) + \mu p_{0,4}(t) \tag{2.1}$$

ii. The server is in WV mode but in busy state $\left(\varsigma(t) = 1\right)$.

$$\frac{dp_{0,1}(t)}{dt} = -\left(\lambda\sigma + b\gamma + \bar{b}\gamma + \theta\right)p_{0,1}(t) + \lambda\sigma p_{0,0}(t) \tag{2.2}$$

$$\frac{dp_{n,1}(t)}{dt} = -\left(\lambda\sigma(1 - \delta_{n,k}) + b\gamma + \bar{b}\gamma + \theta\right)p_{n,1}(t) + \lambda\sigma p_{n-1,1}(t), \quad 1 \le n \le k \tag{2.3}$$

iii. The server is rendering additional service in WV mode $\left(\varsigma(t) = 2\right)$.

$$\frac{dp_{n,2}(t)}{dt} = -\xi p_{n,2}(t) + \bar{b}\gamma p_{n,1}(t) \tag{2.4}$$

iv. The server being in normal busy (NB) mode but in free state $\left(\varsigma(t) = 3\right)$.

$$\frac{dp_{1,3}(t)}{dt} = -(\alpha + \lambda)p_{1,3}(t) + b\gamma p_{1,1}(t) + \xi p_{1,2}(t) + \mu p_{1,4}(t), \quad 0 \le n \le k \tag{2.5}$$

$$\frac{dp_{n,3}(t)}{dt} = -(n\alpha + \lambda)p_{n,3}(t) + b\gamma p_{n,1}(t) + \xi p_{n,2}(t) + \mu p_{n,4}(t), \quad 1 \le n \le k \tag{2.6}$$

v. The server is rendering service during NB mode $\left(\left(\varsigma(t) = 4\right)\right)$.

$$\frac{dp_{0,4}(t)}{dt} = -(\lambda\sigma + \mu)p_{0,4}(t) + \theta p_{0,1}(t) + \alpha p_{1,3}(t) \tag{2.7}$$

$$\frac{dp_{n,4}(t)}{dt} = -(\lambda\sigma(1 - \delta_{n,k}) + \mu)p_{n,4}(t) + \theta p_{n,1}(t) + \lambda\sigma p_{n-1,4}(t), \quad 1 \le n \le k - 1 \tag{2.8}$$

$$\frac{dp_{k,4}(t)}{dt} = -\mu p_{k,4}(t) + \theta p_{k,1}(t) + \lambda\sigma p_{k-1,4}(t) \tag{2.9}$$

where, $\delta_{n,k}$ is the Kronecker delta.

The set of ordinary differential Eqs. (2.1–2.9) are solved by using numerical method viz. IV order Runge-Kutta (R-K) technique. It is noticed that R-K technique can be easily implemented using software like Mathematica, Maple, Matlab, etc. We shall obtain the transient probabilities using routine "ode45" in Matlab. The key performance measures can be derived using the transient probabilities of system states.

2.4   PERFORMANCE MEASURES

The key performance metrics namely average queue length, probabilities for different states and throughput at time t are derived in terms of transient probabilities as follows:

i. The mean queue length at time t is

$$E[N(t)] = p_{0,0}(t) + \sum_{n=0}^{k}\sum_{i=1}^{2} np_{n,i}(t) + \sum_{n=1}^{k} np_{n,3}(t) + \sum_{n=0}^{k} np_{n,4}(t) \tag{2.10}$$

ii. The transient probability for the server being on in WV mode and rendering primary service is

$$P_{WV}(t) = \sum_{n=0}^{k} p_{n,1}(t) \tag{2.11}$$

iii. The transient probability for the server being in WV mode and rendering additional service is

$$P_{IM}(t) = \sum_{n=0}^{k} p_{n,2}(t) \tag{2.12}$$

iv. The transient probability for the server is in NB mode but in free state

$$P_{NF}(t) = \sum_{n=1}^{k} p_{n,3}(t) \tag{2.13}$$

v. The transient probabilities that the server is busy in rendering service in NB mode

$$P_{NB}(t) = \sum_{n=0}^{k} p_{n,4}(t) \tag{2.14}$$

vi. Throughput at time t is given by

$$Th(t) = \left(\mu \sum_{n=0}^{k} p_{n,4}(t) + \gamma \sum_{n=0}^{k} p_{n,2}(t) \right) \tag{2.15}$$

2.5   COST ANALYSIS

The main focus of this study is to frame the cost function $TC(t)$ with consideration of different cost elements and activities. The cost per unit time is defined by

$$TC(t) = C_{WV} P_{WV}(t) + C_{IM} P_{IM}(t) + C_{NF} P_{NF}(t) + C_{NB} P_{NB}(t) + C_W E[N(t)] \tag{2.16}$$

where various cost elements per unit time used are

C_{WV} : cost incurred when the server is in WV mode and rendering primary service
C_{IM} : cost incurred when the server is in WV mode and rendering additional service
C_{NF} : cost incurred when the server is in NB mode but in free state
C_{NB} : cost incurred when the server is in NB mode but in busy state
C_W : waiting cost of each customers present in the queue

TABLE 2.1

Performance Indices by Varying Parameters σ and γ

σ	γ	$P_{WV}(t)$	$P_{IM}(t)$	$P_{NF}(t)$	$P_{NB}(t)$	$TC(t)$
	1	0.6163	0.0352	0.2287	0.2529	46.63
0.5	3	0.6558	0.0293	0.1605	0.1726	34.93
	5	0.7039	0.0321	0.1230	0.1310	32.61
	1	0.4715	0.0399	0.3208	0.3558	52.98
0.7	3	0.5000	0.0307	0.2401	0.2566	36.16
	5	0.5551	0.0340	0.1927	0.2027	32.81
	1	0.3595	0.0404	0.3867	0.4427	56.80
0.9	3	0.3707	0.0290	0.3034	0.3317	36.42
	5	0.4238	0.0323	0.2530	0.2704	32.32

2.6   SENSITIVITY ANALYSIS

By taking illustration, we obtain numerical results for the expected queue length, throughput, etc. The computer program is coded using MATLAB software to compute the performance measures and exploring the sensitivity of the system descriptors. Following are the default parameters which we will use for computing the performance metrics:

$$\alpha = 1,\ \mu = 3,\ \lambda = 1.5,\ \gamma = 2,\ \xi = 4,\ \theta = 0.3,\ \sigma = 0.6\ \text{and}\ b = 0.6$$

$$C_{WV} = \$30,\ C_{IM} = \$40,\ C_{NF} = \$20,\ C_{NB} = \$50,\ C_W = \$10$$

From Table 2.1, for the fixed value of balking probability σ, we notice the decreasing trends in P_{NF}, P_{NB} and increasing trend in P_{WV} with the increments in the primary service rate (γ) during WV. For fix value of γ, with the increase in balking probability (σ), increasing trends in P_{NF}, P_{NB} and decreasing trend in P_{WV} can also observed. From Table 2.2, for a fixed value of σ during WV, we notice the increasing trends in

TABLE 2.2

Performance Indices by Varying Parameters σ and λ

σ	λ	$P_{WV}(t)$	$P_{IM}(t)$	$P_{NF}(t)$	$P_{NB}(t)$	TC
	1.5	0.8127	0.0607	0.0768	0.0497	33.41
0.5	1	0.6596	0.0293	0.1929	0.1334	31.62
	0.5	0.5483	0.0840	0.2194	0.1493	30.70
	1.5	0.7483	0.0399	0.2922	0.3558	33.96
0.7	1	0.5571	0.0841	0.3208	0.3683	31.92
	0.5	0.4163	0.1192	0.4467	0.2045	30.88
	1.5	0.6848	0.0553	0.1626	0.0984	33.72
0.9	1	0.4679	0.1054	0.3939	0.2876	31.94
	0.5	0.3211	0.1420	0.4612	0.3892	30.99

TABLE 2.3
Performance Indices by Varying Parameters σ and α

σ	α	$P_{WV}(t)$	$P_{IM}(t)$	$P_{NF}(t)$	$P_{NB}(t)$	$TC(t)$
	1	0.7919	0.3069	0.0724	0.1033	36.09
0.5	3	0.8085	0.0700	0.0413	0.0958	33.41
	5	0.8159	0.0705	0.0279	0.0907	32.57
	1	0.6869	0.0805	0.1021	0.1453	36.44
0.7	3	0.7117	0.0824	0.0625	0.1405	33.96
	5	0.7241	0.0835	0.0441	0.1360	33.09
	1	0.5824	0.0851	0.1212	0.1769	35.22
0.9	3	0.6127	0.0880	0.0789	0.1786	33.72
	5	0.7184	0.1004	0.0406	0.1162	33.13

P_{NF}, P_{NB} and decreasing trend in P_{WV} with the increments in the arrival rate (λ) during WV. From Table 2.3, we notice that for fix value of α by enhancing balking probability (σ), probability that the server is in normal busy period and the server is free (P_{NF}), probability that the server is in normal busy mode and busy (P_{NB}) increase but P_{WV} decreases. From Table 2.4, for a fixed value of σ, we see the decreasing trend in P_{WV} while enhancing the value of probability (b) during WV. In Tables 2.1–2.4, for a fixed σ, we have seen that there are minor changes in $P_{IM}(t)$ for varying values of parameters γ, λ, α, b, respectively.

In Figure 2.2(i–iv), mean queue length $E[N(t)]$ builds up as time t passes; however, initially, it changes slightly, but as time grows, a remarkable increment is noticed. From Figure 2.2(ii and iv), we observe that as arrival rate (λ) and probability (b) increase, the queue length $E[N(t)]$ also increases. On the other hand, when we increase the service rates γ and μ, there is significant decrement in the queue length as clearly noticed in Figure 2.2(i and iii). From Figure 2.2(v), after

TABLE 2.4
Performance Indices by Varying Parameters σ and b

σ	b	$P_{WV}(t)$	$P_{IM}(t)$	$P_{NF}(t)$	$P_{NB}(t)$	$TC(t)$
	0.4	0.8833	0.0325	0.0605	0.0584	36.09
0.5	0.6	0.8537	0.0631	0.0486	0.0495	37.27
	0.8	0.8344	0.0922	0.0330	0.0381	38.51
	0.4	0.8705	0.0866	0.0118	0.0250	36.44
0.7	0.6	0.8533	0.0737	0.0293	0.0381	38.06
	0.8	0.8492	0.0410	0.0562	0.0579	39.81
	0.4	0.7722	0.0474	0.0885	0.0853	35.22
0.9	0.6	0.7114	0.1314	0.0604	0.0638	37.00
	0.8	0.7051	0.1307	0.0633	0.0661	38.95

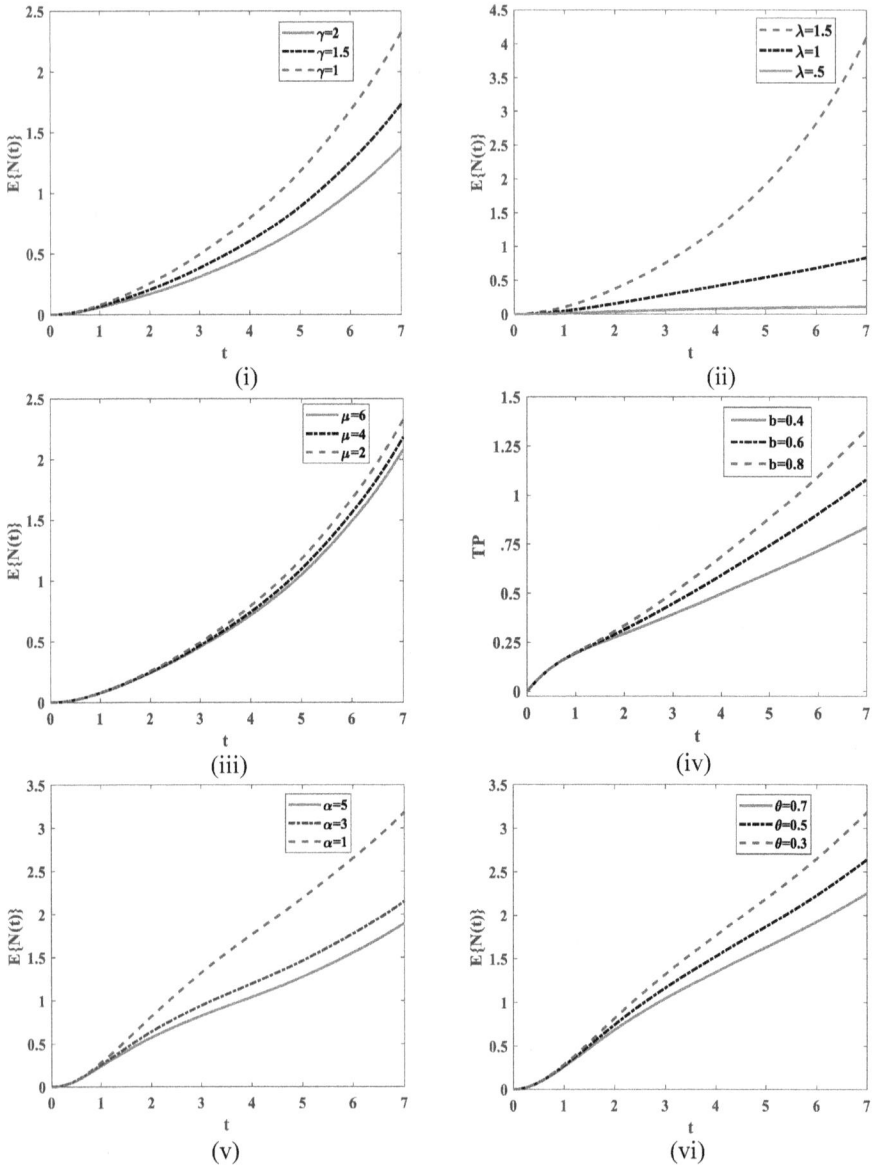

FIGURE 2.2 Expected queue size $E\{N(t)\}$ with variation in (i) γ (ii) λ (iii) μ (iv) b (v) α (vi) θ.

time $t = 2.5$, the queue length $\left(E\{N(t)\}\right)$ seems to lower down with the increasing value of retrial rate (α). From Figure 2.2(vi), we observe that until the time $t = 2$ reaches, the average queue length reveals very less effect for any variation in the value of θ, but later on there is gradually decrement for increasing value of θ.

The significant effects of time and different parameters on throughput are noticed in Figure 2.3(i–iv). From Figure 2.3(i), we observe that beyond time $t = 2.4$, the throughput decreases significantly by giving increment in γ. It is clear from Figure 2.3(ii) that the throughput (TP) quickly increases by the increasing arrival rate (λ). Also, for lower value of arrival rate (say $\lambda = 0.5$), throughput is almost constant. During normal state, as service rate (μ) increases, the throughput seems to decrease. The system throughput (TP) also decreases with the increment in the probability b, which is shown in Figure 2.3(iv). Figure 2.4 portrays the cost function that demonstrates the increasing trend as time grows up. As service rate of server in WV mode increases, the cost decreases but effect becomes more prevalent as time passes.

In Figure 2.5, trends of the server being in different states by varying time have been depicted. Based on numerical results, we can infer that by including the service during WV, the cost can be controlled to a prespecified level, and the objective of better grade of service (GoS) can be achieved.

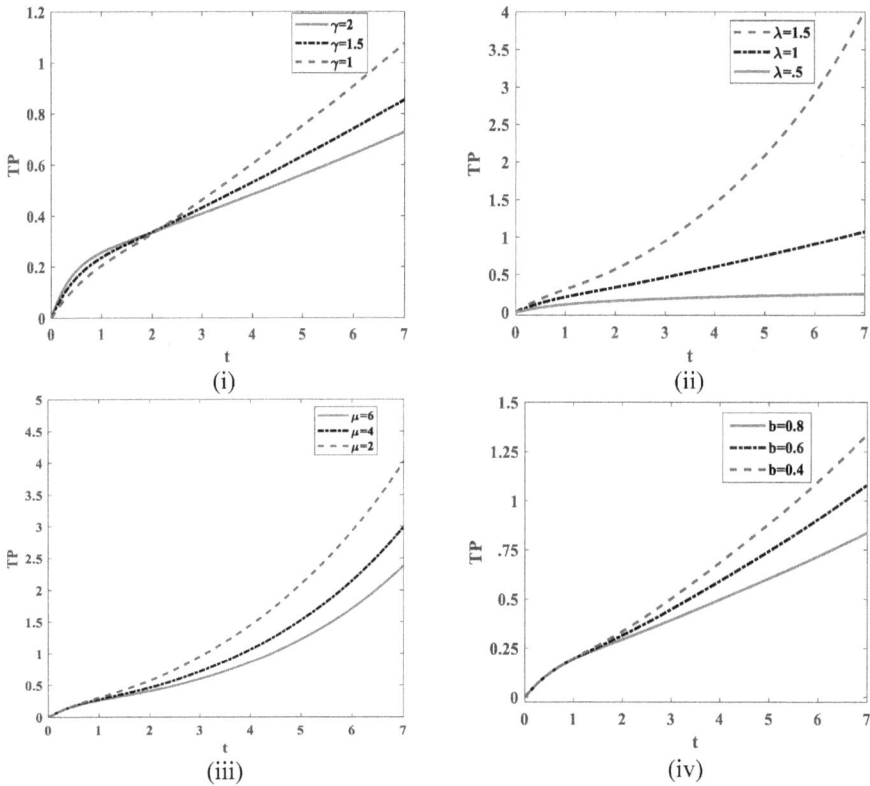

FIGURE 2.3   Throughput (TP) with variation in (i) γ (ii) λ (iii) μ (iv) b.

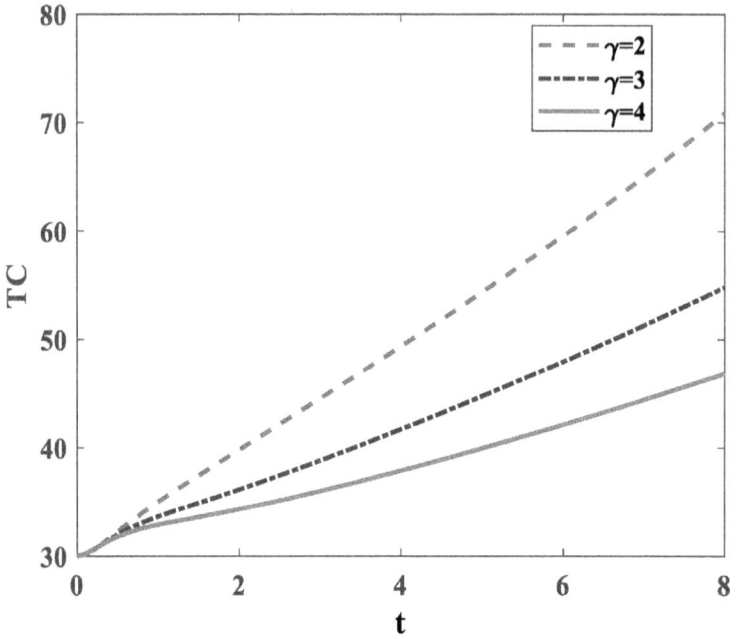

FIGURE 2.4   Total cost (TC) vs. t.

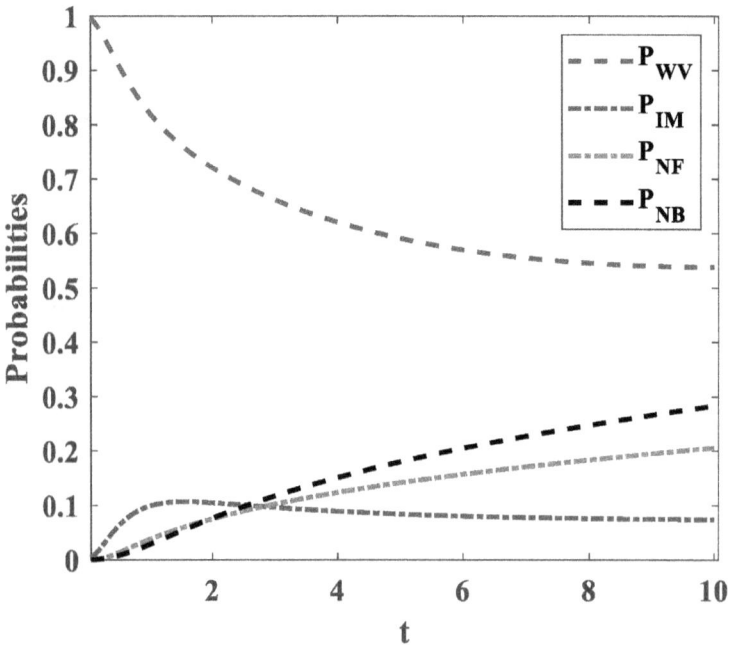

FIGURE 2.5   Probabilities for different system states.

2.7 CONCLUSION

The main focus of this research work is to facilitate the performance indices of the transient model of retrial queue with imperfect service during WV and balking behaviour of the customers. In many applications, including malls, doctor's clinics, call centres, communication networks, e-ticketing systems, etc., the server can provide service in WV mode at a slower pace, and as such the customers may not be satisfied and may require additional service. The inclusion of realistic features of balking, retrial attempts and option of additional service during the WV period allow our model to deal with real-world congestion problem. The model developed in the general setup can be further modified by including the concepts of unreliable server, optimal control strategy viz. N-policy and/or F-policy, etc.

REFERENCES

1. Jain, M. and Bhagat, A. 2015. Transient analysis of finite F-policy retrial queues with delayed repair and threshold recovery. *National Academy Science Letters*. 38: 257–261, https://doi.org/10.1007/s40009-014-0337-1.
2. Sudhesh, R., Azhagappan, A. and Dharmaraja, S. 2017. Transient analysis of M/M/1 queue with working vacation, heterogeneous service and customers' impatience. *RAIRO – Operations Research*. 5: 591–606, https://doi.org/10.1051/ro/2016046.
3. Azhagappan, A., Veeramani, E., Monica, W. and Sonabharathi, K. 2018. Transient solution of an M/M/1 retrial queue with reneging from orbit. *Applications and Applied Mathematics: An International Journal*. 13: 628–638.
4. Ammar S.I. 2015. Transient analysis of an M/M/1 queue with impatient behavior and multiple vacations. *Applied Mathematics and Computation*. 260: 97–105.
5. Servi, L.D. and Finn, S.G. 2002. M/M/1 queues with working vacations (M/M/1/WV). *Performance Evaluation*. 50: 41–52, https://doi.org/10.1016/S0166-5316(02)00057-3.
6. Ezeagu, N.J., Orwa, G.O. and Winckler, M.J. 2018. Transient analysis of a finite capacity M/M/1 queuing system with working breakdowns and recovery policies. *Global Journal of Pure and Applied Mathathematics*. 14: 1049–1065.
7. Jain, M., Rani, S. and Singh, M. 2019. Transient analysis of Markov feedback queue with working vacation and discouragement. *Performance Prediction and Analytics of Fuzzy, Reliability and Queuing Models*. 235–250, https://doi.org/10.1007/978-981-13-0857-4_18.
8. Sudhesh, R. and Azhagappan, A. 2018. Transient analysis of an M/M/1 queue with variant impatient behavior and working vacations. *OPSEARCH*. 55: 787–806, https://doi.org/10.1007/s12597-018-0339-8.
9. Vijayashree, K.V. and Janani, B. 2018. Transient analysis of an M/M/1 queueing system subject to differentiated vacations. *Quality Technology and Quantity Management*. 15: 730–748, https://doi.org/10.1080/16843703.2017.1335492.
10. Kumar, R. and Sharma, S. 2019. Transient analysis of an M/M/c queuing system with retention of reneging customers. *International Journal of Operational Ressearch*. 36: 78–91, https://doi.org/10.1504/IJOR.2019.102071.
11. Legros, B. 2019. Transient analysis of a Markovian queue with deterministic rejection. *Operations Research Letters*. 47: 391–397, https://doi.org/10.1016/j.orl.2019.07.005.
12. Krishna Kumar, B., Krishnamoorthy, A., Madheswari, S.P. and Sadiq B.S. 2007. Transient analysis of a single server queue with catastrophes, failures and repairs. *Queueing Systems*. 56: 133–141, https://doi.org/10.1007/s11134-007-9014-0.

13. Azhagappan, A. and Deepa, T. 2017. Transient analysis of N-policy queue with system disaster repair preventive maintenance re-service balking closedown and setup times. *Journal of Industrial & Management Optimization*. 13: 1–14, https://doi.org/10.3934/jimo.2019083.

14. Azhagappan, A. and Deepa, T. 2019. Transient analysis of a Markovian single vacation feedback queue with an interrupted closedown time and control of admission during vacation. *Applications & Applied Mathematics*. 14: 34–45.

15. Jain, M. and Rani, S. 2020. Performance analysis of Markovian retrial queue with unreliable server by incorporating the features of balking and reneging. *Proceedings of National Academy of Science India, Section A Physical Science*. 1–8: published online first, https://doi.org/10.1007/s40010-020-00667-z.

3 Finite Capacity Tandem Queueing Network with Reneging

Chandra Shekhar
Birla Institute of Technology and Science, Pilani,
Rajasthan, India

Neeraj Kumar, Amit Gupta
Uttarakhand Technical University, Dehradun, India

Rajesh Kumar Tiwari
Nimbus Academy of Management, Dehradun, India

CONTENTS

3.1   INTRODUCTION

In the current age of IoT, cloud computing, industry 4.0, etc., modern computer networks process heavy workloads with random service demands. Hence, computer networks face challenges such as traffic control, time delay, data loss, or the blocking of a new connection. For the predictive analysis, various performance measures, including throughput, loss probability, blocking and delay, can be easily evaluated by using the queueing theoretic approach. The queueing models deal with waiting line problems that are surprisingly prevalent in all spheres of real-life problems, since waiting lines are ubiquitous. Jobs arrive at a node (router) in the computer network, wherein jobs have to processed and forwarded by the node. A node processes one job at a time and places the remaining jobs in the waiting line (buffer) of finite capacity. Queueing network is also used to study the quality of service at various stages of the computer network.

The Markov process is widely used for the performance analysis of various queueing problems that arise in any service system, including a computer network. A Markov process consists of a set of states and a labeled transition between the states. A state of the Markov process represents distinct conditions of interest in the computer network, namely various types of waiting jobs, the number of failed services, and so on.

In the tandem queueing network, several service facilities in a network are connected in a series to provide the requested service to the job sequentially. In the tandem queue, blocking or starving is a common phenomenon, and buffer size is an important decision criterion for optimal services in the network. Reneging, an unpredictable impatience behavior, is a vital aspect of a tandem queueing network. In the reneging process, the job leaves the waiting queue without being served. For example, in service call centers, jobs may hang up before completing their task.

Queueing theory is a mathematical tool that is commonly applied in the design and development of computer networks. Queueing modeling has been used effectively for evaluating the performance measures of computer networks, and many researchers in the past enrich literature. Some important works on the queueing network have been done by Perros [1994], Bolch et al. [1998], Altman [2000], Balsamo et al. [2003], Bylina and Bylina [2005], etc. Filipowicz and Kwiecien [2008] discussed different queueing networks for the modeling of complex network systems and presented formula that can be used to improve the flow of data or packets, throughput, delay time, etc. A lot of literature and textbooks concerning the queueing theory are available [Gross and Harris, 2008]. Petrovic et al. [2008] presented an application of the Markov theory to study the model of networked transport system in the view of determining the system performance. Jain et al. [2009] developed a single and batch service finite capacity queueing model for the communication system, and presented the transient analysis of the average queue length, expected idle period, and expected busy period. Cominetti and Guzman [2013] presented a multipath routing network model that combines the two network models, i.e., NUM (Network Utility Maximization) and MTE (Markovian Traffic Equilibrium), and explored the congestion control and multipath routing. Shekhar and Jain [2013] dealt with a Markovian queueing system having multitask service counters that provide complete service in three stages in tandem with the state-dependent rate and having a finite queue in front of each counter. Czachorski [2015] focused on three types of approaches, namely Markov models, fluid flow approximation, and diffusion approximation, to analyze transient states in queueing models. Baumann and Sandmann [2017] proposed a finite buffer multiserver tandem queue system with distributed stage-type service time and discussed various performance measures. Quality of service (QoS) enhances the performance of computer networks through proper management of available resources. Delay is a significant measure of QoS. Guan et al. [2006] developed a queueing model with delay constraints and presented an algorithm on queue with the moveable threshold that controls the delay. Ahmad et al. [2009] presented a comprehensive analysis of various congestion control schemes focused on certain performance measures such as throughput, mean queue length, probability of packet loss and end-to-end delay. Ammar [2017] discussed a Markovian model with impatient jobs and evaluated the performance of the system. Wang and Zhang [2018] developed a queueing model with impatient clients within the multiserver queueing system along with the threshold structure and discussed the various performance measures. Wang et al. [2019] introduced a two-station tandem queueing system with reneging and abandonment and described numerous performance metrics of a tandem.

This chapter considers the Markovian model that provides three types of service requests with two nodes. Both nodes with finite buffer are modeled as single-server finite capacity queues, and jobs make independent reneging decisions while waiting in the second node. Our study is applicable in the architectural process involved in routers where multitype jobs like data, packets, voice, etc. require processing at node(s) in sequence. The purpose of this chapter is threefold. First, we develop a mathematical model for the real-time, nodes-jobs computing network in tandem architect. Second, we derive the steady-state solution for the given model and obtain the different performance measures of the computing network. Third, we construct a relationship between the costs to determine the optimal size of the buffer in the rendering of the service. Numerical simulations have also been made for the direct benefit to the decision makers. The chapter is structured in the following manner. In section 3.2, we present a detailed model description of the tandem queueing network and formulate governing Chapman-Kolmogorov differential-difference equation. In section 3.3, we define performance measures to validate our governing model. In section 3.4, we develop the cost function of the computing network for optimal analysis of decision parameters. The results of the numerical experiment have been tabulated and depicted in section 3.5. Conclusions are drawn in section 3.6 along with future scopes.

3.2 MODEL DESCRIPTION

We consider a tandem queue network with two nodes: Node 1 and Node 2, which are arranged in series with a dedicated buffer of finite size. Each node is assumed as a single-server queueing system. Three types of service requests are processed within the tandem network. In type 1 service requests, jobs require the processing only at the first node and leave the network after first node service completion. In type 2 service requests, jobs require processing at both nodes in sequence. In type 3 service requests, jobs require only the second node's service. Jobs may make independent reneging decisions while waiting in the buffer space for the second node. The architectural layout is depicted in Figure 3.1. For the mathematical modeling purpose, we consider the following assumptions and notations.

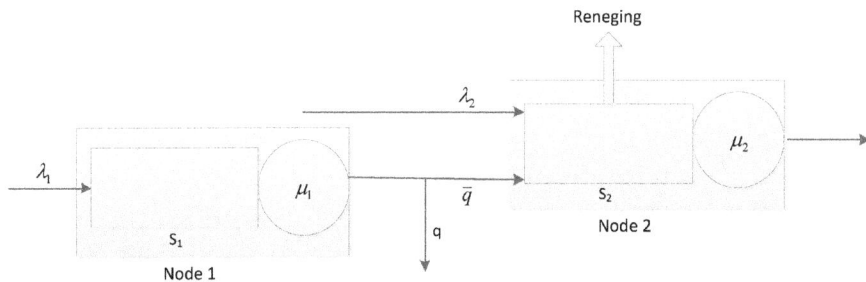

FIGURE 3.1   The tandem queue network with reneging.

At Node 1, jobs arrive at the computing network following a Poisson process with a rate λ_1. On arrival, if the server of Node 1 is available idly, the job enters for service without waiting. Otherwise, the job waits in Node 1's queue in the buffer of size S_1. The server at Node 1 renders the service whose service time follows an exponential distribution with mean rate μ_1. After a job's processing request is satisfied at Node 1, the job may enter in Node 2 with probability $\bar{q}$ if it requires Node 2 service, otherwise it leaves the computing network. Those jobs that require Node 2's service only enter in Node 2 directly with exponentially distributed interarrival time with rate λ_2. The server at Node 2 provides the service to both types of jobs, considering that they are waiting in one queue in the buffer of size S_2. The service time at Node 2 follows the exponential distribution with mean rate μ_2. Jobs may make independent reneging decisions while waiting at Node 2. Job reneging follows the Poisson process with the mean rate γ.

Using the concept of the Markov process, the state transition diagram is illustrated in Figure 3.2. Let $\{I(t); t \geq 0\}$ and $\{J(t); t \geq 0\}$ be the number of jobs in the

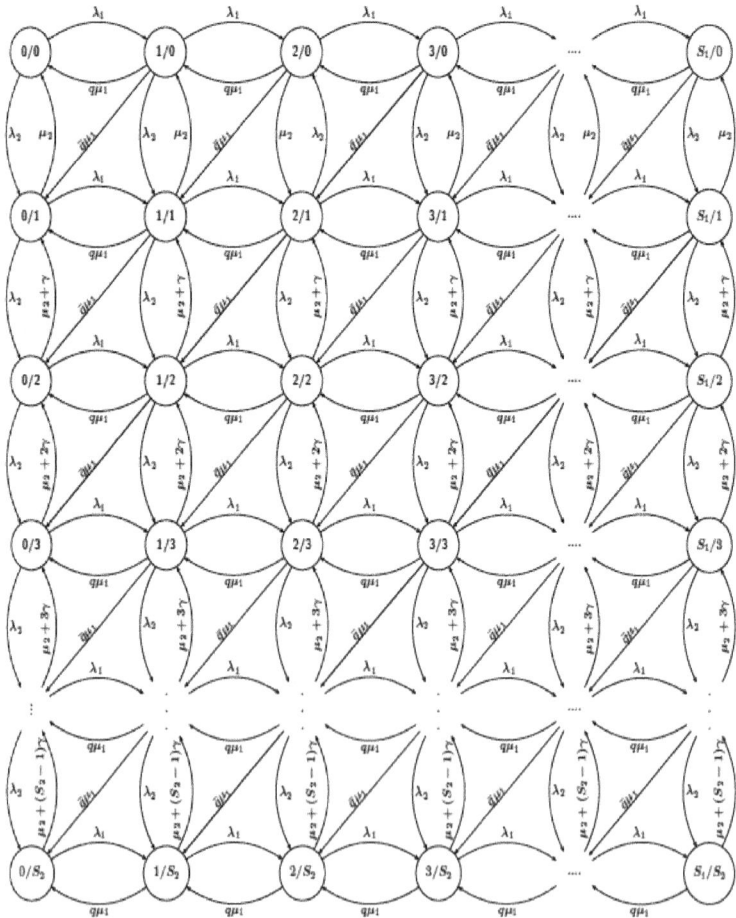

FIGURE 3.2   State transition diagram.

queue of Node 1 and Node 2, respectively, at the time t. Then, $\{I(t), J(t); t \geq 0\}$ is a continuous Markov process on the state space $S = \{(i, j); 0 \leq i \leq S_1, 0 \leq j \leq S_2\}$. Let $P_{i,j}(t) = \text{Prob}\{I(t) = i, J(t) = j; t \geq 0\}$ be the probability that there are i jobs in the queue at Node 1, j jobs at Node 2 at the time t.

For the steady-state analysis, we define $P_{i,j} = \lim_{t \to \infty} P_{i,j}(t)$. Then, the set of forward Chapman-Kolmogorov difference equations for the above model is as follows:

$$-(\lambda_1 + \lambda_2)P_{0,0} + q\mu_1 P_{1,0} + \mu_2 P_{0,1} = 0 \tag{3.1}$$

$$-(\lambda_1 + \lambda_2 + \mu_1)P_{i,0} + \lambda_1 P_{i-1,0} + q\mu_1 P_{i+1,0} + \mu_2 P_{i,1} = 0; 1 \leq i \leq S_1 - 1 \tag{3.2}$$

$$-(\lambda_2 + \mu_1)P_{S_1,0} + \lambda_1 P_{S_1-1,0} + \mu_2 P_{S_1,1} = 0 \tag{3.3}$$

$$-(\lambda_1 + \lambda_2 + (\mu_2 + (j-1)\gamma))P_{0,j} + \lambda_2 P_{0,j-1} + q\mu_1 P_{1,j} + \bar{q}\mu_1 P_{1,j-1}$$
$$+ (\mu_2 + j\gamma)P_{0,j+1} = 0; \quad 1 \leq j \leq S_1 - 1 \tag{3.4}$$

$$-\left(\lambda_1 + \lambda_2 + \mu_1 + (\mu_2 + (j-1)\gamma)\right)P_{i,j} + \lambda_1 P_{i-1,j} + \lambda_2 P_{i,j-1} + q\mu_1 P_{i+1,j}$$
$$+ \bar{q}\mu_1 P_{i+1,j-1} + (\mu_2 + j\gamma)P_{i,j+1} = 0; \quad 1 \leq i \leq S_1 - 1; 1 \leq j \leq S_2 - 1 \tag{3.5}$$

$$-(\lambda_2 + (\mu_2 + (S_2 - 1)\gamma))P_{S_1,j} + \lambda_1 P_{S_1-1,j} + \lambda_2 P_{S_1,j-1}$$
$$+ (\mu_2 + j\gamma)P_{S_1,j+1} = 0; \quad 1 \leq j \leq S_2 - 1 \tag{3.6}$$

$$-(\lambda_1 + (\mu_2 + (S_2 - 1)\gamma))P_{0,S_2} + \lambda_2 P_{0,S_2-1} + q\mu_1 P_{1,S_2} + \bar{q}\mu_1 P_{1,S_2-1} = 0 \tag{3.7}$$

$$-\left(\lambda_1 + q\mu_1 + (\mu_2 + (S_2 - 1)\gamma)\right)P_{i,S_2} + \lambda_1 P_{i-1,S_2} + \lambda_2 P_{i,S_2-1} + q\mu_1 P_{i+1,S_2}$$
$$+ \bar{q}\mu_1 P_{i+1,S_2-1} = 0; \quad 1 \leq i \leq S_1 - 1 \tag{3.8}$$

$$-(q\mu_1 + (\mu_2 + (S_2 - 1)\gamma))P_{S_1,S_2} + \lambda_1 P_{S_1-1,S_2} + \lambda_2 P_{S_1,S_2-1} = 0 \tag{3.9}$$

For the computational convenience, Eqs. (3.1–3.9) can be represented in the matrix form, as:

$$\mathbf{XP} = \mathbf{0} \tag{3.10}$$

where $\mathbf{X}$ is the coefficient matrix of order $S_1 \times S_2$, $\mathbf{P}$ is the probability vector of state probabilities and $\mathbf{0}$ is the null vector of size $S_1 \times S_2$. Using the normalizing condition of probability

$$\mathbf{Pe} = \mathbf{1} \tag{3.11}$$

where $\mathbf{e}$ denotes the unit row vector of order $S_1 \times S_2$, Eq. (3.10) can be expressed as

$$\mathbf{X_R P = Y} \tag{3.12}$$

where $\mathbf{X_R}$ is derived from the matrix $\mathbf{X}$ simply by replacing the last row with a row vector with all unit elements and $\mathbf{Y} = [0,0,0,\ldots\ldots,0,1]^T$ is a column vector of order $S_1 \times S_2$. Non-homogenous system of equations Eq. (3.12) has been solved by a numerical technique that is based on SOR (Successive-Over-Relaxation) method, which is an extrapolation to the Gauss-Seidel method by taking the relaxation parameter $w = 1.25 (1 < w \le 2)$ to improve the convergence rate [Jain et al. 2011].

3.3 PERFORMANCE MEASURES

In this section, using the steady-state probabilities derived in the previous section, we develop following performance measures of tandem queueing system:

1. The average number of jobs at Node 1 and Node 2, respectively

$$E[N_1] = \sum_{i=0}^{S_i} \sum_{j=0}^{S_j} i P_{i,j} \tag{3.13}$$

$$E[N_2] = \sum_{i=0}^{S_i} \sum_{j=0}^{S_j} j P_{i,j} \tag{3.14}$$

2. Expected waiting time of jobs at Node 1 and Node 2, respectively

$$E[W_1] = E[N_1] / \lambda_{eff_1} \tag{3.15}$$

$$E[W_2] = E[N_2] / \lambda_{eff_2} \tag{3.16}$$

where effective arrival rate of jobs at Node 1 and Node 2, respectively, are obtained as

$$\lambda_{eff_1} = \sum_{i=0}^{S_1-1} \sum_{j=0}^{S_2} \lambda_1 P_{i,j} \quad \text{and} \quad \lambda_{eff_2} = \sum_{i=0}^{S_1} \sum_{j=0}^{S_2-1} (\lambda_2 + \bar{q}\mu_1) P_{i,j}.$$

3. Probability that the server at respective Node 1 and Node 2 is busy

$$UT_1 = 1 - \sum_{j=1}^{S_2} P_{0,j} \tag{3.17}$$

$$UT_2 = 1 - \sum_{i=1}^{S_1} P_{i,0} \tag{3.18}$$

4. Probability of the lost job at the Node 1 and Node 2, respectively

$$PS_1 = \sum_{j=0}^{S_2} P_{S_1,j} \tag{3.19}$$

$$PS_2 = \sum_{i=0}^{S_1} P_{i,S_2} \tag{3.20}$$

5. Effective reneging rate

$$RR = \sum_{i=0}^{S_1} \sum_{j=2}^{S_2} (j-1)\gamma P_{i,j} \tag{3.21}$$

6. Throughput of the computing network

$$\tau = \sum_{i=0}^{S_1} \sum_{j=1}^{S_2} \mu_2 P_{i,j} \tag{3.22}$$

3.4   COST FUNCTION

For the optimal analysis, we require some objective function (expected cost function) in terms of decision parameters. To develop the expected cost function for the computing network, the following unit costs of different states are assumed:

C_1: Holding cost per unit time of each waiting job at Node 1
C_2: Holding cost per unit time of each waiting job at Node 2
C_3: Service cost per unit time of each waiting job at Node 1
C_4: Service cost per unit time of each waiting job at Node 2
C_5: cost per unit time of the waiting buffer space at Node 1
C_6: cost per unit time of the waiting buffer space at Node 2

Hence, the expected total cost is

$$E(TC) = C_1 E(N_1) + C_2 E(N_2) + C_3 \mu_1 + C_4 \mu_2 + C_5 S_1 + C_6 S_2 \tag{3.23}$$

3.5   NUMERICAL RESULTS

Performance measures described in the previous section are computed numerically through the MATLAB program for providing quick insights about the model to decision makers. The expected waiting times and expected number of the jobs in the network and other described performance measures are calculated for different parameters and summarized in tables and graphs. For numerical experimentation,

TABLE 3.1

Performance Measures by Varying S_1 and λ_1

S_1	λ_1	$E[W_1]$	$E[W_2]$	UT_1	UT_2	PS_1	PS_2	RR
	0.8	4.7675	0.7250	0.7982	0.5166	2.33E-03	4.31E-05	8.52E-02
20	1.2	15.5420	0.9216	0.9956	0.5868	1.70E-01	1.58E-04	1.22E-01
	1.6	18.3392	0.9267	1.0000	0.5883	3.75E-01	1.65E-04	1.23E-01
	0.8	4.9638	0.7267	0.7998	0.5172	2.48E-04	4.41E-05	8.55E-02
30	1.2	25.1373	0.9259	0.9993	0.5880	1.67E-01	1.64E-04	1.23E-01
	1.6	28.3391	0.9268	1.0000	0.5883	3.75E-01	1.65E-04	1.23E-01
	0.8	4.9957	0.7268	0.8000	0.5172	2.66E-05	4.42E-05	8.55E-02
40	1.2	35.0402	0.9266	0.9999	0.5882	1.67E-01	1.65E-04	1.23E-01
	1.6	38.3406	0.9268	1.0000	0.5883	3.75E-01	1.65E-04	1.23E-01

we fix the default governing parameters of studied computing network as follows: $S_1 = 30$; $S_2 = 10$; $q = 0.1$; $\lambda_1 = 0.6$; $\lambda_2 = 0.4$; $\mu_1 = 1$; $\mu_2 = 2$; $\gamma = 0.2$.

In Table 3.1, the variation of the different performance measures is tabulated for different arrival rate λ_1 from 0.8 to 1.6 for different buffer capacity (S_1). A similar type of depiction for the variation of the performance measures with the arrival rate λ_2 for different buffer capacity (S_2) of Node 2 is done in Table 3.2.

Tables 3.1 and 3.2 reveal that expected waiting time at Node 1 and Node 2 increases with the arrival rate of jobs and buffer size. The apparent results validate our modeling and analysis. Both tables also summarize the possibility that servers at respective Node 1 and Node 2 are busier when jobs are joining the computing network at a faster rate. The probability of lost jobs decreases with the increased size of the buffer in the computing network. This result prompts decision makers to decide buffer size optimally based on cost analysis. The reneging of jobs is more vulnerable when there are more jobs in waiting due to a faster rate.

TABLE 3.2

Performance Measures by Varying S_2 and λ_2

S_2	λ_2	$E[W_1]$	$E[W_2]$	UT_1	UT_2	PS_1	PS_2	RR
5	0.8	2.6706	0.7196	0.6151	0.6003	1.87E-07	2.40E-02	1.19E-01
	1.2	2.9946	0.8846	0.6404	0.7293	6.47E-07	6.24E-02	2.03E-01
	1.6	3.6117	1.0383	0.6806	0.8246	3.88E-06	1.19E-01	2.92E-01
10	0.8	2.5014	0.7462	0.6001	0.6034	8.90E-08	2.20E-04	1.33E-01
	1.2	2.5129	0.9680	0.6012	0.7399	9.37E-08	1.98E-03	2.58E-01
	1.6	2.5646	1.2184	0.6058	0.8449	1.19E-07	9.52E-03	4.34E-01
15	0.8	2.5000	0.7465	0.6000	0.6034	8.84E-08	5.77E-07	1.33E-01
	1.2	2.5001	0.9713	0.6000	0.7400	8.85E-08	1.91E-05	2.60E-01
	1.6	2.5016	1.2350	0.6001	0.8457	8.91E-08	2.53E-04	4.48E-01

TABLE 3.3
Performance Measures by Varying γ and μ_1

γ	μ_1	$E[W_1]$	$E[W_2]$	UT_1	UT_2	PS_1	PS_2	RR
0.2	0.5	50.2543	0.7351	0.9993	0.4026	1.67E-01	3.41E-06	4.44E-02
	1.0	2.5000	0.5564	0.6000	0.4419	8.84E-08	8.89E-06	5.63E-02
	1.5	1.1111	0.4133	0.4000	0.4419	6.92E-13	9.03E-06	5.63E-02
0.8	0.5	50.2538	0.5939	0.9993	0.3715	1.67E-01	1.11E-08	1.07E-01
	1.0	2.5000	0.4369	0.6000	0.4046	8.84E-08	2.91E-08	1.31E-01
	1.5	1.1111	0.3246	0.4000	0.4046	6.92E-13	2.92E-08	1.31E-01
1.4	0.5	50.2538	0.5345	0.9993	0.3560	1.67E-01	3.85E-10	1.38E-01
	1.0	2.5000	0.3891	0.6000	0.3864	8.84E-08	1.01E-09	1.67E-01
	1.5	1.1111	0.2890	0.4000	0.3864	6.92E-13	1.01E-09	1.67E-01

Tables 3.3 and 3.4 depict the variation of performance measures by varying γ and μ_1 and by changing q and μ_2, respectively. Tables 3.3 and 3.4 show how service rates μ_1, μ_2 govern the computing network efficiently. The result prompts that the service rates μ_1, μ_2 are also the decision parameters for the network design. The reneging rate γ must be low to increase the revenue from the computing network.

In Figure 3.3(i–iv), we depict the variation of critical performance measure, the expected number of jobs in the computing network with respect to arrival rates of the job at Node 1. It is apparent that $E(N)$ is an increasing function of λ_1. It is highly influenced by the buffer size of Node 1(S_1), as S_1 increases $E(N)$ increase, but other factors like queue size of Node 2 (S_2) leave out probability (q) and the reneging rate (γ) have a light impact on $E(N)$. In Figure 3.3(v–viii), we present the variation of $E(N)$ vs λ_2 where $E(N)$ has less implication for S_1 and S_2, and significant decrement is observed as q and γ increase. Figure 3.4 shows a similar

TABLE 3.4
Performance Measures by Varying q and μ_2

q	μ_2	$E[W_1]$	$E[W_2]$	UT_1	UT_2	PS_1	PS_2	RR
0.1	0.7	2.5105	1.9708	0.6010	0.8553	9.27E-08	1.55E-03	3.41E-01
	1.1	2.5017	1.2163	0.6002	0.6930	8.91E-08	2.60E-04	1.78E-01
	1.5	2.5003	0.8175	0.6000	0.5596	8.85E-08	5.17E-05	1.01E-01
0.5	0.7	2.5005	1.8531	0.6000	0.7328	8.86E-08	1.45E-04	1.87E-01
	1.1	2.5001	1.1205	0.6000	0.5537	8.85E-08	1.94E-05	9.10E-02
	1.5	2.5000	0.7630	0.6000	0.4328	8.84E-08	3.44E-06	5.08E-02
0.9	0.7	2.5000	1.8688	0.6000	0.5462	8.84E-08	3.65E-06	7.76E-02
	1.1	2.5000	1.1296	0.6000	0.3856	8.84E-08	3.98E-07	3.58E-02
	1.5	2.5000	0.7859	0.6000	0.2934	8.84E-08	6.40E-08	1.99E-02

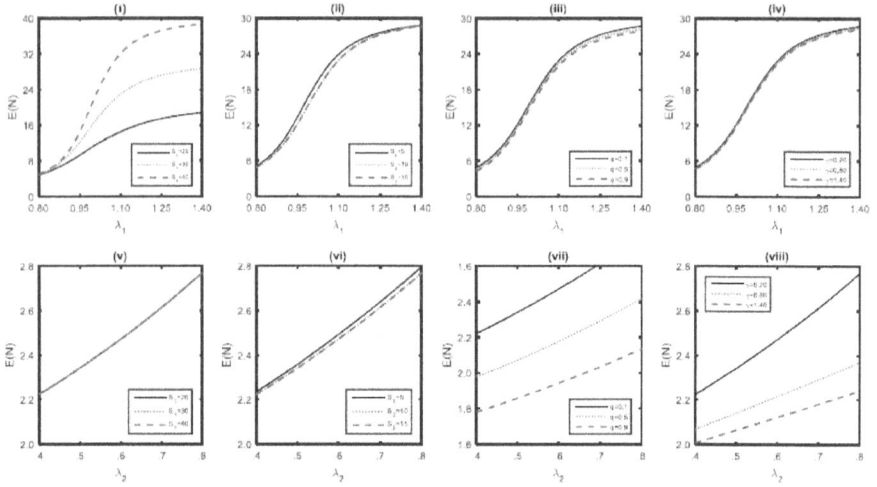

FIGURE 3.3 Expected number of jobs in the computing network wrt arrival rates of jobs.

contrast for service rates, wherein $E(N)$ decreases rapidly with an increment of μ_1 and μ_2 as expected.

Figures 3.5 and 3.6 depict the change in the value of the throughput of the computing network with respect to arrival rates and service rates, respectively, for different network parameters. It is observed that the throughput of the computing network (τ) is less affected by buffer sizes S_1 and S_2. In contrast, as the value of q and γ increase, the throughput of the computing network τ decreases significantly.

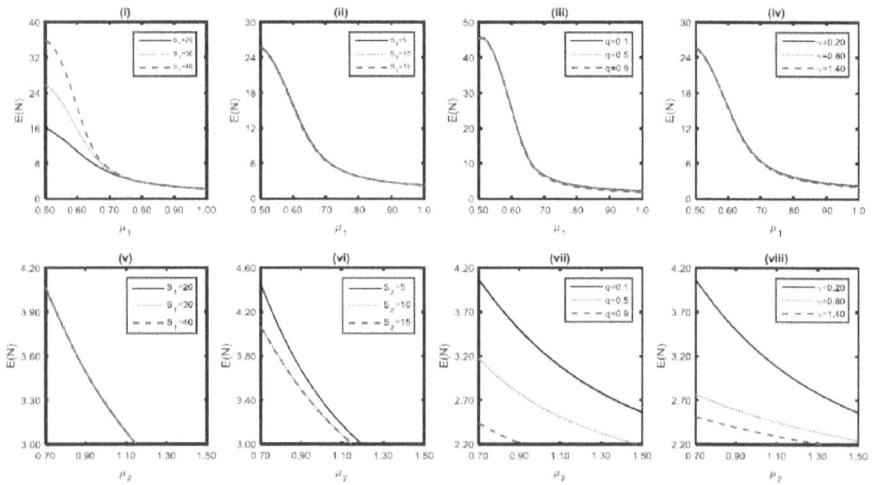

FIGURE 3.4 Expected number of jobs in the computing network wrt service rates of jobs.

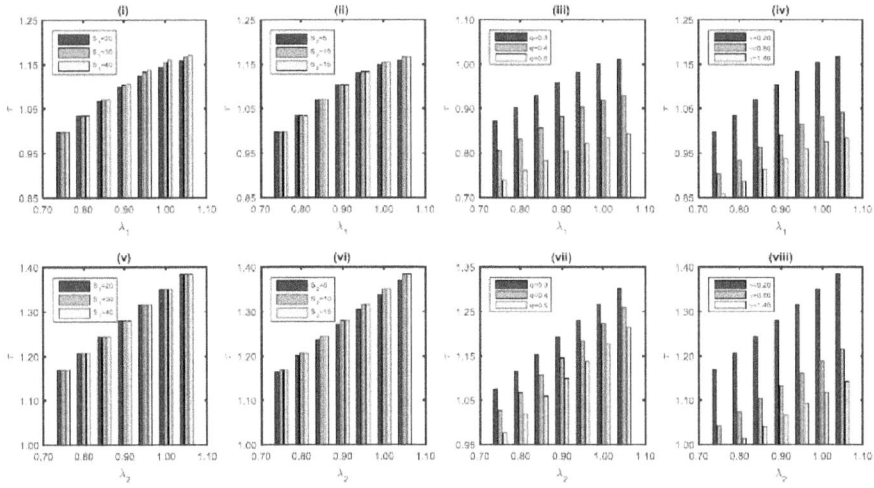

FIGURE 3.5 Throughput of the computing network wrt arrival rates of jobs.

For computing the expected total cost, we fix the default value of the unit costs as follows: $C_1 = 40$; $C_2 = 60$; $C_3 = 2$; $C_4 = 3$; $C_5 = 10$; $C_6 = 12$. Variation with respect to governing system parameters is depicted in Figures 3.7–3.9. Figures 3.7 and 3.8 indicate that the expected total cost $\left(E(TC)\right)$ increases as S_1 and S_2 increase, but $E(TC)$ decreases when q and γ increase. Figure 3.9 indicates that $E(TC)$ is a convex function for parameters μ_1 and μ_2, which prompts that the service rates suitable decision parameters for the optimal analysis.

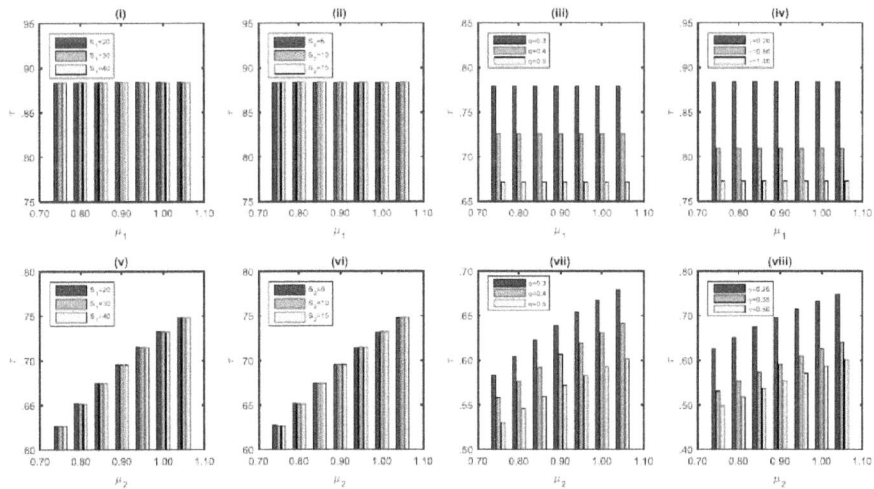

FIGURE 3.6 Throughput of the computing network wrt service rates of jobs.

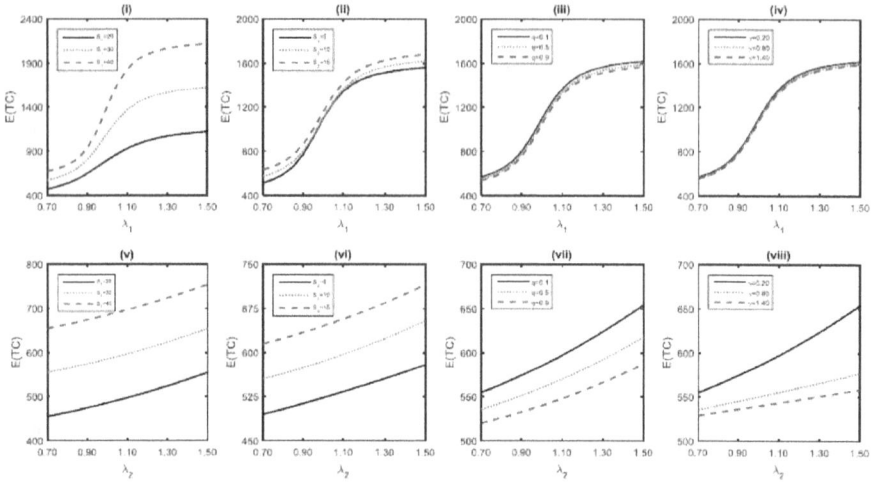

FIGURE 3.7   Expected total cost of the computing network wrt arrival rates of jobs.

For the efficient working of the computing network, the analyst must be judgmental for deciding the buffer size at nodes and service rates for jobs at nodes to avoid the reneging of the arrived jobs.

3.6   CONCLUSION

The queueing theory has been used effectively for evaluating the performance measures of the computing network. This chapter presents the performance analysis of a state-dependent tandem computing network with a dedicated server at each node.

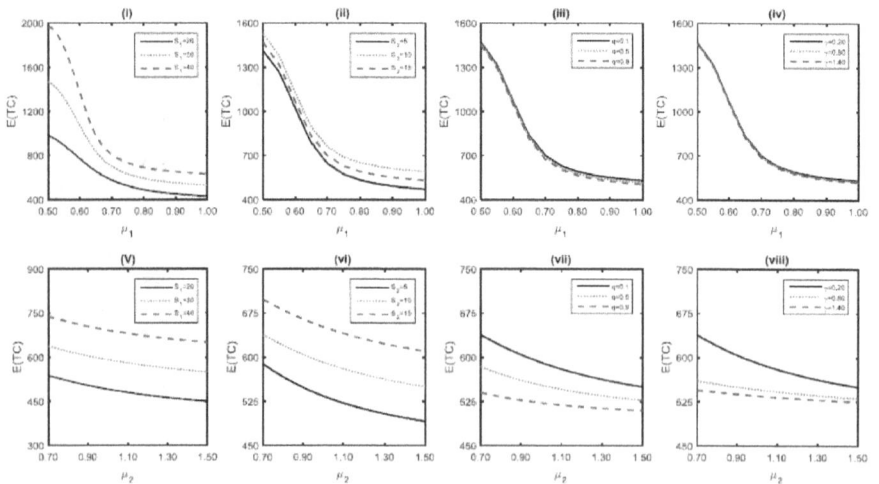

FIGURE 3.8   Expected total cost of the computing network wrt service rates of jobs.

(i)

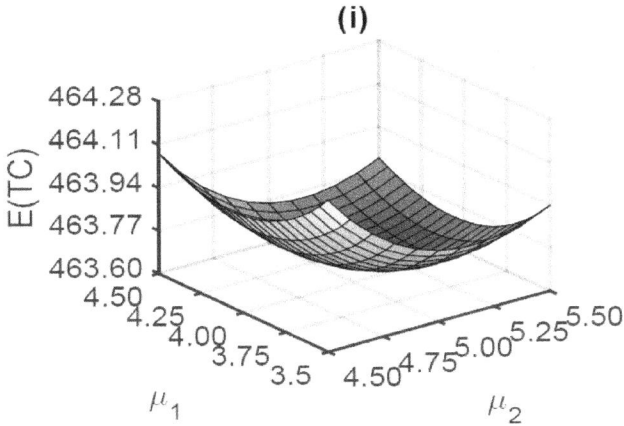

FIGURE 3.9 Convex nature of the expected total cost of the computing network wrt service rates of jobs.

Using the advanced numerical procedure, we compute steady-state probabilities of the computing network and some important performance measures with regard to the expected number of waiting jobs in the computing network, expected waiting time, network cost, etc. Results show that the Markov process plays a vital role in the performance analysis of various queueing problems that arise in any service system, including computer networks. This study is helpful for the network analyst to build a real-life, reliable and cost-effective network.

REFERENCES

Ahmad, S., Mustafa, A., Ahmad, B., Bano, A. and Hosam, A. 2009. Comparative study of congestion control techniques in high-speed networks. *International Journal of Computer Science and Information Security*. 6: 222–231.

Altman, E. 2000. Applications of Markov decision processes in communication network: A survey (Research Report). RR-3984. INRIA. France. 1-51 inria-00072663.

Ammar, S.I. 2017. Transient solution of an *M/M/*1 vacation queue with a waiting server and impatient customers. *Journal of the Egyptian Mathematical Society*. 25: 337–342.

Balsamo, S., Persone, V.D.N. and Inverardi, P. 2003. A review on queueing network models with finite capacity queues for software architectures performance prediction. *Performance Evaluation*. 51: 269–288.

Baumann, H. and Sandmann, W. 2017. Multi-server tandem queue with Markovian arrival process, phase-type service times, and finite buffers. *European Journal of Operation Research*. 256: 187–195.

Bolch, G., Greiner, S., DeMeer, H. and Trivedi, K.S. 1998. *Queueing Networks and Markov Chains: Modelling and Performance Evaluation with Computer Science Applications*. Second ed. Hoboken, NJ: Wiley.

Bylina, B. and Bylina, J. 2005. Using Markov chains for modeling networks. *Annals UMCS Informatica AI 3, Poland*. 3: 27–34.

Cominetti, R. and Guzman, C. 2013. Network congestion control with Markovian multipath routing. *Mathematical Programming, Springer*. 147: 231–251.

Czachorski, T. 2015. Queueing models for performance evaluation of computer networks–transient state analysis. *Springer Proceedings in Mathematics & Statistics.* 116: 51–80.

Filipowicz, B. and Kwiecien, J. 2008. Queueing systems and networks: Models and applications. *Bulletin of the Polish Academy of Sciences, Technical Sciences.* 56: 379–390.

Gross, D. and Harris, C.M. 2008. *Fundamentals of Queueing Theory.* Fourth Ed. Hoboken, NJ: Wiley.

Guan, L., Woodward, M.E. and Awan, I.U. 2006. Control of queueing delay in a buffer with time-varying arrival rate. *Journal of Computer and System Sciences.* 72: 1238–1248.

Jain, M., Sharma, G.C., Saraswat, V.K. and Rakhee, R. 2009. Transient analysis of a telecommunication system using state dependent Markovian queue under bi-level control policy. *Journal of King Abdulaziz University, Engineering Science.* 20: 77–90.

Jain, M., Sharma, G.C. and Sharma, R. 2011. Working vacation queue with service interruption and multi-optional repair. *International Journal of Information Management Science.* 22: 157–175.

Perros, H. G. 1994. *Queueing Networks with Blocking: Exact and Approximate Solutions.* New York, NY: Oxford University Press.

Petrovic, G., Petrovic, N. and Marinkovic, Z. 2008. Application of the Markov theory to queueing networks. *Journal Facta Universitatis, Series Mechanical Engineering.* 6: 45–56.

Shekhar, C. and Jain, M. 2013. Finite queueing model with multi-task servers and blocking. *American Journal of Operational Research.* 3: 17–25.

Wang, J., Abouee-Mehrizi, H., Baron, O. and Berman, O. 2019. Tandem queues with impatient customers. *Performance Evaluation.* 135: 102011.

Wang, Q. and Zhang, B. 2018. Analysis of a busy period queueing system with balking, reneging and motivating. *Applied Mathematical Modelling.* 64: 480–488.

4 Transient Solution of Markov Model for Fault Tolerant System with Redundancy

K.P.S. Baghel
Rajkiya Mahavidhyalaya Targawan Jaithra, Etah, India

Madhu Jain
Indian Institute of Technology Roorkee, Uttarakhand, India

CONTENTS

4.1 INTRODUCTION

For the up gradation of reliability and performance of a machining system, the provision of fault tolerance measures, namely maintenance as well as redundancy during the design and operating stages, is required. The inventory of spares has been made available in many real-time machining systems to increase the reliability/availability and system capability as well as to produce the desired output. From the system designer's point of view, the knowledge of optimal number of spares and repairmen, operative problems and scheduling of the machines can be helpful to achieve the mission reliability. The life-time and repair attributes of the faulty components should be examined so as to predict the performance metrics for the better design and grade of service of the concerned machining system operating in fault tolerance environment. Therefore, the system engineer attempts to find the size of repair crew and standby units to keep the fault tolerant machining system (FTMS) in working operation despite of some components' failures and can achieve the desired level of production.

The machine repair system (MRS) has the physical limitation as far as waiting space is concerned. When a queue of machines waiting for repair reaches to a certain length, one can stop the joining of failed machines in the MRS until a waiting space is made available by the repair-completion. When a machine stops while working in the system, it causes the loss of production. Therefore, to maintain the continuity/regularity of the production, the necessity of redundancy as well as maintenance is felt. The repairmen can help in removing and repairing of the faulty units as and when required so that the same can be further used. Many queue theorists and practitioners have contributed their valuable research works in this direction (cf., Baker, 1973, Gaver and Lehoezky, 1977; Chiang and Niu, 1981; Goel et al., 1985; Agnihothri, 1989; Gupta, 1995; Moustafa, 1998; Wang and Ke, 2000; Jain et al., 2004; Chikara et al., 2006; Jain et al., 2009; Shinde et al., 2011; Jain, 2013; Jain et al., 2016; Jain and Gupta, 2018; Kumar et al., 2019). Jain et al. (2020) investigated performance metrics of FTMS which can operate in spite of unavailability of server due to service interruption causes by the vacations and breakdowns.

When the in-flow of the failed machines and repair rates are influenced by the status of the repair station and already present workload in the system, the state-dependent rates can be used to describe the queueing model of machining system. Some efforts by eminent queue theorists in this direction have also been made (cf. Jain, 1997a, b; Jain and Baghel, 2001; Jain et al., 2008; Jain and Chauhan, 2010; Singh et al., 2013). The broken down machines demanding repair may balk and/ or renege without getting repair due to heavy workload or other reason. Some of the eminent queue theorists who studied the machine repair problems with state dependent rates are Blackburn (1972), Al-seedy (1992), Jain and Lata (1994), Jain et al. (2000, 2004), Singh and Jain (2007), Jain and Mittal (2008), Maheshwari et al. (2010), Jain et al. (2014), Dhakad and Jain (2016), and Sharma et al. (2017). The state dependent queueing system operating under control F-policy was studied by Jain and Sanga (2019). They have discussed the specialized scenarios concerned with machine repair system and time-sharing system so as to explore the sensitiveness of the system descriptors with respect to indices established.

The main objective of this chapter is to develop Markov queueing model for FTMS having provision of standbys and multi-repairmen by considering realistic features, namely state dependent rates. The description of the model is presented in section 4.2. The governing equations for transient states and their solution by using matrix method are given in section 4.3. Some system indices as well as cost function are obtained in section 4.4. Section 4.5 facilitates the sensitivity analysis to demonstrate the significant impact of key system descriptors on the performance indices. In section 4.6, concluding remarks are given.

4.2 MATHEMATICAL FORMULATION

Consider the fault tolerant system having a group of Q operating and W cold standby machines. We denote the total machines in the FTMS by $T = Q + W$. The FTMS has repair crew having s repairmen. It is assumed that the life time and repair time of failed machines follow Markov property. The service facility provides repair according to FCFS basis. The failure rates of operating machines are equal to λ, while μ is the repair rate of each failed machine. The constant parameters a and b show the degree of the reneging and balking functions, respectively. Two different scenarios can arise.

4.2.1 CASE I: $s \leq W$

The effective failure rates and repair rates for the system are considered as follows:

$$\Lambda_j = \begin{cases} Q\lambda, & 0 < j < s \\[2ex] Q\left(\dfrac{s}{j+1}\right)^b \lambda, & s \leq j < W \\[2ex] (T-j)\left(\dfrac{s}{j+1}\right)^b \lambda, & W \leq j < T \end{cases} \tag{4.1}$$

and

$$\mu = \begin{cases} j\mu, & 0 < j \leq s \\[2ex] \left(\dfrac{j}{s}\right)^a s\mu, & s < n \leq T \end{cases} \tag{4.2}$$

The governing equations for different transient states are obtained as:

$$\frac{d\pi_0(t)}{dt} = -Q\lambda\,\pi_0(t) + \mu\pi_1(t) \tag{4.3}$$

$$\frac{d\pi_j(t)}{dt} = -[Q\lambda + j\mu]\,\pi_j(t) + Q\lambda\pi_{j-1}(t) + (j+1)\mu\pi_{j+1}(t),\ 1 \leq j\ < s \tag{4.4}$$

$$\frac{d\pi_s(t)}{dt} = -\left[Q\left(\frac{s}{s+1}\right)^b \lambda + s\mu\right]\pi_s(t) + Q\lambda\pi_{s-1}(t) + \left(\frac{s+1}{s}\right)^a s\mu\pi_{s+1}(t) \tag{4.5}$$

$$\frac{d\pi_j(t)}{dt} = -\left[Q\left(\frac{s}{j+1}\right)^b \lambda + \left(\frac{j}{s}\right)^a s\mu\right]\pi_j(t) + Q\lambda\left(\frac{s}{j}\right)^b \pi_{s-1}(t)$$
$$+ \left(\frac{j+1}{s}\right)^a s\mu\,\pi_{j+1}(t),\ s < j < W \tag{4.6}$$

$$\frac{d\pi_W(t)}{dt} = -\left[(T-W)\left(\frac{s}{W+1}\right)^b \lambda + \left(\frac{W}{s}\right)^a s\mu\right]\pi_W(t) + Q\lambda\left(\frac{s}{W}\right)^b \pi_{W-1}(t)$$
$$+ \left(\frac{W+1}{s}\right)^a s\mu\,\pi_{W+1}(t) \tag{4.7}$$

$$\frac{d\pi_j(t)}{dt} = -\left[(T-j)\left(\frac{s}{j+1}\right)^b \lambda + \left(\frac{j}{s}\right)^a s\mu\right]\pi_j(t) + (T-j+1)\lambda\left(\frac{s}{j}\right)^b \pi_{j-1}(t)$$

$$+ \left(\frac{j+1}{s}\right)^a s\mu\,\pi_{j+1}(t), W < j < T \tag{4.8}$$

$$\frac{d\pi_T(t)}{dt} = -\left(\frac{T}{s}\right)^a s\mu\,\pi_T(t) + \lambda\left(\frac{s}{T}\right)^b \pi_{T-1}(t) \tag{4.9}$$

4.2.2   CASE II: $W < s$

In this case, the birth and death rates are given by:

$$\Lambda_j = \begin{cases} Q\lambda, & 0 < j < W \\ (T-j)\left(\dfrac{s}{j+1}\right)^b \lambda, & W \leq j < T \end{cases} \tag{4.10}$$

and

$$\mu_j = \begin{cases} j\mu, & 0 < j \leq s \\ \left(\dfrac{j}{s}\right)^a s\mu, & s < n \leq T \end{cases} \tag{4.11}$$

In this case, the governing equations are given by:

$$\frac{d\pi_0(t)}{dt} = -Q\lambda\,\pi_0(t) + \mu\,\pi_1(t) \tag{4.12}$$

$$\frac{d\pi_j(t)}{dt} = -[Q\lambda + j\mu]\,\pi_j(t) + Q\lambda\,\pi_{j-1}(t) + (j+1)\,\mu\,\pi_{j+1}(t), \; 1 < j < W \tag{4.13}$$

$$\frac{d\pi_W(t)}{dt} = -\left[(T-W)\left(\frac{s}{s+1}\right)^b \lambda + s\mu\right]\pi_W(t) + Q\lambda\pi_{W-1}(t) + s\mu\,\pi_{W+1}(t) \tag{4.14}$$

$$\frac{d\pi_j(t)}{dt} = -\left[Q\left(\frac{s}{j+1}\right)^b \lambda + j\mu\right]\pi_j(t) + Q\lambda\left(\frac{s}{j}\right)^b \pi_{j-1}(t) + (s+1)\mu\,\pi_{j+1}(t), \; W < j < s \tag{4.15}$$

$$\frac{d\pi_s(t)}{dt} = -\left[(T-s)\left(\frac{s}{s+1}\right)^b \lambda + s\mu\right]\pi_s(t) + (T-s+1)\lambda\pi_{s-1}(t) + \left(\frac{s+1}{s}\right)^a s\mu\,\pi_{s+1}(t) \tag{4.16}$$

$$\frac{d\pi_j(t)}{dt} = -\left[(T-j)\left(\frac{s}{j+1}\right)^b \lambda + \left(\frac{j}{s}\right)^a s\mu\right]\pi_j(t) + (T-j+1)\lambda\left(\frac{s}{j}\right)^b \pi_{j-1}(t)$$

$$+\left(\frac{j+1}{s}\right)^a s\mu\, \pi_{j+1}(t), \quad s < j < T \tag{4.17}$$

$$\frac{d\pi_T(t)}{dt} = -\left(\frac{T}{s}\right)^a s\mu\, \pi_T(t) + \lambda\left(\frac{s}{T}\right)^b \pi_{T-1}(t) \tag{4.18}$$

4.2.3 THE LIMITING CASE

When t tends to infinity, i.e., for the steady state we denote the state probabilities by $\pi_j = \underset{t\to\infty}{\text{Lim}}\,\pi_j(t)$. In this case, using recursive approach, the above Eqs. (4.3)–(4.9) and (4.12)–(4.18) provide the explicit product form solution as follows:

$$\pi_j = \pi_0 \prod_{i=0}^{j-1} \frac{\lambda_i}{\mu_{i+1}}, \quad j = 0, 1, 2,\dots, T \tag{4.19}$$

where

$$\pi_0 = \left[1 + \sum_{j=1}^{T}\left\{\prod_{i=1}^{j-1}\left(\frac{\lambda_i}{\mu_{i+1}}\right)\right\}\right]^{-1}$$

4.3 THE TRANSIENT SOLUTION

Laplace transforms of Eqs. (4.3)–(4.9) can be written in matrix equation by:

$$\mathbf{C}(\alpha)\mathbf{P}(\alpha) = \mathbf{I}_{T+1} \tag{4.20}$$

Here, $\mathbf{C}(\alpha)$ denotes $(T+1)\times(T+1)$ real symmetric tri-diagonal matrix.

$$\text{Also,}\quad \mathbf{P} = [P_0(\alpha),\, P_1(\alpha),\, P_2(\alpha),\dots,P_{s-1}(\alpha), P_s(\alpha),\dots ,P_T(\alpha)]' \tag{4.21}$$

$$\mathbf{I}_{T+1} = [1,0,0,\dots,0,0]' \tag{4.22}$$

Here dash (′) denotes the transpose of the matrix.
    Laplace transform of set of Eqs. (4.12)–(4.18) yield

$$\mathbf{D}(\alpha)\mathbf{Q}(\alpha) = \mathbf{I}_T \tag{4.23}$$

Here $\mathbf{D}(\alpha)$, is $T \times T$ tridiagonal matrix that can be constructed by leaving the first row and first column of $\mathbf{C}(\alpha)$.

$$\text{Also,}\ \mathbf{Q}(\alpha) = [q_1(\alpha),\ q_2(\alpha),\dots,q_s(\alpha),\dots,q_W(\alpha),\dots,q_T(\alpha)]' \tag{4.24}$$

After applying Cramer's rule, we obtain matrices $\mathbf{C}_k(\alpha)$ and $D_k(\alpha)$ from Eqs. (4.20) and (4.23), respectively, by substituting the kth column by the RHS unit vector. Now,

$$p_k(\alpha) = \frac{|C_k(\alpha)|}{|C(\alpha)|}, \qquad 0 \le k \le T \tag{4.25}$$

and

$$q_k(\alpha) = \frac{|D_k(\alpha)|}{|D(\alpha)|}, \qquad 1 \le k \le T \tag{4.26}$$

Since determinants $|C(\alpha)|$ and $|D(\alpha)|$ have real and distinct zeros, we can rewrite $p_k(\alpha)$ and $q_k(\alpha)$ in partial fraction form. Also, we have

$$|C(\alpha)| = \alpha|\gamma(\alpha)| \quad \text{and} \quad |D(\alpha)| = |\delta(\alpha)| \tag{4.27}$$

where

$$\gamma(\alpha) = \begin{bmatrix}
\begin{bmatrix} \alpha+\mu \\ +Q\lambda \end{bmatrix} & -\sqrt{\mu Q\lambda} & 0 & .. & 0 & & 0 \\
-\sqrt{\mu.Q\lambda} & \begin{bmatrix} \alpha+2\mu \\ +Q\lambda \end{bmatrix} & -\sqrt{2\mu.Q\lambda} & .. & 0 & & 0 \\
.. & .. & .. & .. & .. & & .. \\
0 & 0 & 0 & .. & -\sqrt{Q\left(\frac{s}{W+1}\right)^b \lambda.\left(\frac{W}{s}\right)^a s\mu} & & \begin{bmatrix} \alpha+(T-W)\left(\frac{s}{W+1}\right)^b \lambda \\ +\left(\frac{W+1}{s}\right)^a s\mu \end{bmatrix} \\
.. & .. & .. & .. & .. & & .. \\
0 & 0 & 0 & .. & 0 & & 0 \\
 & & 0 & .. & 0 & 0 & 0 \\
 & & 0 & .. & 0 & 0 & 0 \\
 & & .. & .. & .. & .. & .. \\
 & -\sqrt{Q\left(\frac{s}{W+2}\right)^b \lambda.\left(\frac{W+1}{s}\right)^a s\mu} & .. & 0 & 0 & 0 \\
 & & .. & .. & .. & .. & .. \\
 & & 0 & .. & 0 & -\sqrt{\left(\frac{s}{T}\right)^b \lambda.\left(\frac{T-1}{s}\right)^a s\mu} & \begin{bmatrix} \alpha+\left(\frac{s}{T}\right)^b \lambda \\ +\left(\frac{T}{s}\right)^a s\mu \end{bmatrix}
\end{bmatrix}$$

$$\tag{4.28}$$

and

$$\delta(\alpha) = \begin{bmatrix} \begin{bmatrix} \alpha \\ +Q\lambda \end{bmatrix} & -\sqrt{\mu.Q\lambda} & 0 & .. & 0 & 0 \\ -\sqrt{\mu.Q\lambda} & \begin{bmatrix} \alpha+2\mu \\ +Q\lambda \end{bmatrix} & -\sqrt{2\mu.Q\lambda} & .. & 0 & 0 \\ .. & .. & .. & .. & .. & .. \\ 0 & 0 & 0 & .. & -\sqrt{Q\left(\dfrac{s}{W+1}\right)^b \lambda.\left(\dfrac{W}{s}\right)^a s\mu} & \begin{bmatrix} \alpha+(T-W)\left(\dfrac{s}{W+1}\right)^b \lambda \\ +\left(\dfrac{W+1}{s}\right)^a s\mu \end{bmatrix} \\ .. & .. & .. & .. & .. & .. \\ 0 & 0 & 0 & .. & 0 & 0 \\ & & 0 & .. \ 0 & 0 & 0 \\ & & 0 & .. \ 0 & 0 & 0 \\ & & .. & .. \ .. & .. & .. \\ -\sqrt{Q\left(\dfrac{s}{W+2}\right)^b \lambda.\left(\dfrac{W+1}{s}\right)^a s\mu} & & & .. \ 0 & 0 & 0 \\ & & .. & .. \ .. & .. & .. \\ 0 & & & .. \ 0 & -\sqrt{\left(\dfrac{s}{T}\right)^b \lambda.\left(\dfrac{T-1}{s}\right)^a s\mu} & \begin{bmatrix} \alpha+\left(\dfrac{s}{T}\right)^b \lambda \\ +\left(\dfrac{T}{s}\right)^a s\mu \end{bmatrix} \end{bmatrix}$$

$$(4.29)$$

It is noted that the zeros of polynomials $|\gamma(\alpha)|$ and $|\delta(\alpha)|$ are the negative eigen values of $[\gamma(0)]$ and $[\delta(0)]$. Let $\gamma_1, \gamma_2,....,\gamma_T$ and $\delta_1, \delta_2,..., \delta_T$ be the eigen values of $[\gamma(0)]$ and $[\delta(0)]$, respectively. Thus:

$$|C(\alpha)| = \alpha \prod_{l=1}^{T}(\alpha - \gamma_l) \quad \text{and} \quad |D(\alpha)| = \prod_{l=1}^{T}(\alpha - \delta_l) \qquad (4.30)$$

where γ_l and $\delta_l\,(l = 1, 2,...,T)$ are all negatives.
   Hence:

$$p_k(\alpha) = \frac{|C_k(\alpha)|}{\alpha \prod\limits_{k=1}^{T}(\alpha - \gamma_k)}, \quad 0 \le k \le T \qquad (4.31)$$

$$q_k(\alpha) = \frac{|D_k(\alpha)|}{\prod\limits_{k=1}^{T}(\alpha - \delta_k)}, \quad 1 \le k \le T \qquad (4.32)$$

Breaking into partial fractions, the RHS of Eqs. (4.31) and (4.32) yield

$$p_k(\alpha) = \frac{a_{0k}}{\alpha} + \sum_{l=1}^{T} \frac{a_{lk}}{(\alpha - \gamma_l)}, \qquad (4.33)$$

and

$$q_k(\alpha) = \sum_{l=1}^{T} \frac{b_{lk}}{(\alpha - \delta_l)}, \qquad (4.34)$$

where

$$a_{0k} = \frac{|C_k(0)|}{\displaystyle\prod_{l=1}^{T} \gamma_l} \qquad (4.35)$$

$$a_{lk} = \frac{|C_k(-\gamma_l)|}{\gamma_l \displaystyle\prod_{\substack{m=1 \\ m \neq l}}^{T} (\gamma_1 - \gamma_m)} \qquad (4.36)$$

and

$$b_{lk} = \frac{|D_k(-\delta_l)|}{\displaystyle\prod_{i=1 \neq l}^{T} (\delta_1 - \delta_i)} \qquad (4.37)$$

4.4   PERFORMANCE INDICES

The inverse LT of Eqs. (4.33) and (4.34) provide

$$\pi_k(t) = a_{0k} + \sum_{l=1}^{T} a_{lk} e^{\gamma_l t}, \quad 0 \leq k \leq T \qquad (4.38)$$

and

$$q_k(t) = \sum_{l=1}^{T} b_{lk} e^{\delta_l t}, \quad 1 \leq k \leq T \qquad (4.39)$$

We notice that the system reliability in terms of probabilities is

$$R(t) = \sum_{k=0}^{W} \pi_k(t) \qquad (4.40)$$

The mean time to system failure (MTTF) is given as

$$MTTF = \int_0^\infty R(u)du = \sum_{k=1}^Q \int_0^\infty q_k(u)du \qquad (4.41)$$

Some system indices, namely, (i) mean number of operating machines E(Q), (ii) mean number of cold standby machines E(W), (iii) mean number of busy repairmen E(B), (iv) mean number of idle repairmen E(I), and (v) operating utilization OU are obtained by:

$$E(Q) = \sum_{j=W+1}^T (j-W)\pi_j(t) \qquad (4.42)$$

$$E(W) = \sum_{j=o}^W (W-j)\pi_j(t) \qquad (4.43)$$

$$E(B) = s - E(I) \qquad (4.44)$$

$$E(I) = \sum_{j=o}^{s-1} (s-j)\pi_j(t) \qquad (4.45)$$

$$OU = \frac{E(B)}{s} \qquad (4.46)$$

Further, we develop a cost function to obtain the optimum combination of repairmen and spares. The various cost elements per unit time are taken as given below:

- C_H: Expenditure incurred in holding the broken down operating machines.
- C_W: Expenditure incurred in case of broken down standby machines.
- C_B: Expenditure incurred on the busy repairmen.
- C_I: Expenditure incurred on the idle repairmen.

The total expected (profit) cost is obtained as:

$$E\{C(t)\} = C_H E(Q) + C_W E(W) + C_B E(B) + C_I E(I) \qquad (4.47)$$

4.5 SENSITIVITY ANALYSIS

For the sensitivity of key descriptors on the system indices, we compute numerical results, which have been shown in Table 4.1 and Figures 4.1–4.8. For computation, we set the default parameter values as $Q = M = 7$, $W = S = 5$, $s = R = 3$, $\lambda = 0.2$, $\mu = 1$, $(a, b) = (0.2, 0.4)$, $C_H = 12$, $C_W = 16$, $C_B = 20$, $C_I = 30$.

TABLE 4.1

Various Performance Measures by Varying W for $Q = 20$ and $(a,\ b) = (0,0)$

W	λ	$E(Q)$	$E(W)$	$E(B)$	$E(C)$
4	0.1	2.83	2.26	1.61	119.51
	0.3	7.75	0.24	2.90	129.26
	0.5	12.41	0.01	3.00	213.61
	0.7	15.82	0.00	3.00	281.46
6	0.1	2.83	4.18	1.61	169.74
	0.3	8.09	0.78	2.90	118.29
	0.5	13.33	0.04	3.00	193.13
	0.7	17.12	0.00	3.00	267.44
8	0.1	2.83	6.17	1.61	224.79
	0.3	8.28	1.74	2.90	124.01
	0.5	14.08	0.14	3.00	172.10
	0.7	18.32	0.00	3.00	251.23

In Table 4.1, we demonstrate the results for the mean number of broken down machines $E(Q)$, mean cold standby machines $E(W)$, mean number of busy repairmen $E(B)$ and average cost $E(C)$ by varying the number of spares (W) and failure rate (λ) for parameters $(a,b) = (0,0)$. We notice that as λ increases, $E(W)$ decreases, but $E(Q)$ and $E(B)$ increase. We also see that as W attains higher value, $E(Q)$, $E(W)$, $E(B)$ and $E(C)$ ramp up. Figure 4.1(a–c) demonstrates the effect of λ on the reliability for $(a,b) = (0.5, 0.3)$, $(a,b) = (1,1)$ and $(a,b) = (1.5, 1.5)$, respectively. We note that when the failure rate (λ) increases, the reliability decreases. As time passes, the reliability seems to lower down in the starting but later on it becomes constant. For particular values of λ, the reliability has the lower value for $(a,b) = (0.5, 0.3)$ in comparison to $(a,b) = (1,1)$ and $(a,b) = (1.5, 1.5)$. It is observed that the reliability for $(a,b) = (1,1)$ is higher than that in the case $(a,b) = (1.5, 1.5)$.

The impacts of increments in operating units on the total cost CT(t) for (a,b) = (0,0), (0.5, 0.3) and (1.5, 1.5) are shown in Figure 4.2(a–c), respectively. We see that as the number of operating machines and time "t" increase, the reliability decreases. For constant values of number of operating machines and t, the reliability reveals the increasing trend.

Figure 4.3(a) and (b) shows the sensitivity of standby units on the reliability for $(a,b) = (0.5, 0.3)$ and $(a,b) = (1,1)$, respectively. We notice that the reliability takes higher values as standby units increase but decrease as time grows up. Furthermore, reliability for $(a,b) = (0.5, 0.3)$ is larger than that for $(a,b) = (1,1)$. Figure 4.3(c) displays the effect of spare part on reliability when the parameters $(a,b) = (1.5, 1.5)$. We note that the reliability for $W = S = 2$ is almost constant, i.e., 1. We also see that as standby units increase, the reliability decreases, but as time increases, reliability diminishes and finally takes almost constant value as time t grows up.

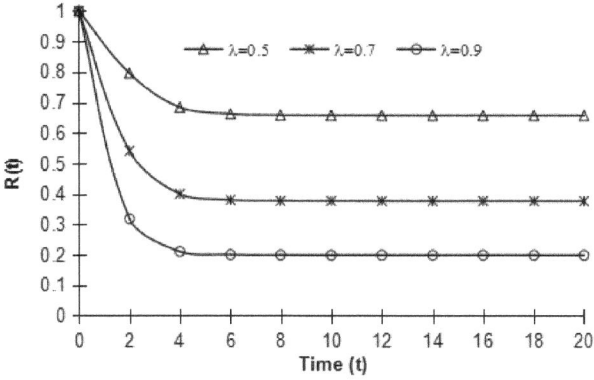

(a)

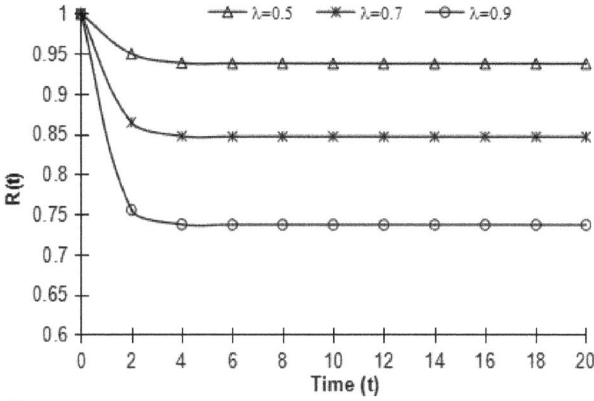

(b)

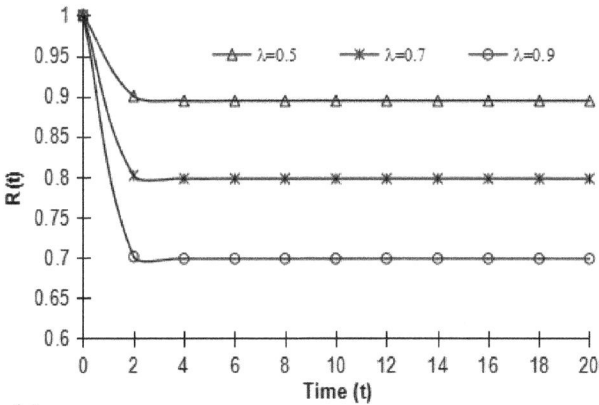

(c)

FIGURE 4.1   (a) Effect of λ on the reliability for $(a,b) = (0.5, 0.3)$; (b) effect of λ on the reliability for $(a,b) = (1, 1)$; (c) effect of λ on the reliability for $(a,b) = (1.5, 1.5)$.

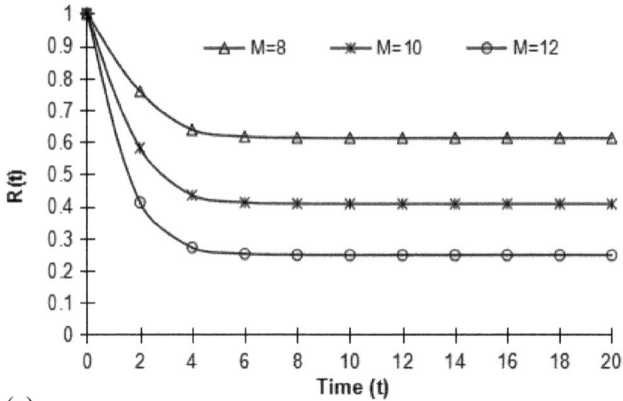

(a)

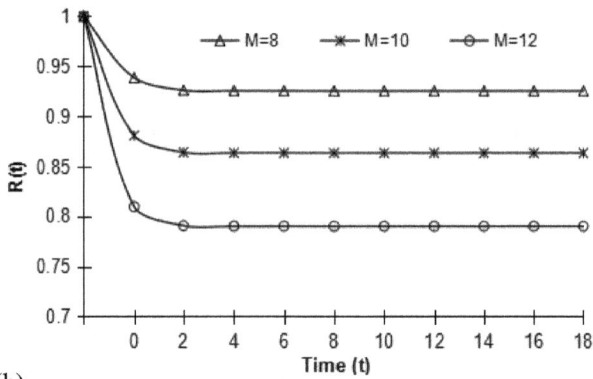

(b)

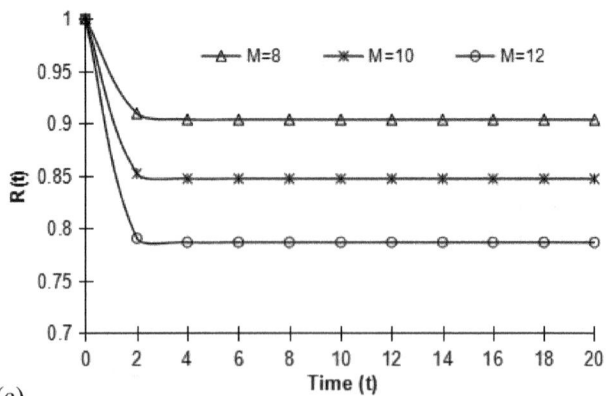

(c)

FIGURE 4.2   (a) Effect of number of operating machines on the reliability for $(a,b) = (0.5, 0.3)$; (b) effect of number of operating machines on the reliability for $(a,b) = (1, 1)$; (c) effect of number of operating machines on the reliability for $(a,b) = (1.5, 1.5)$.

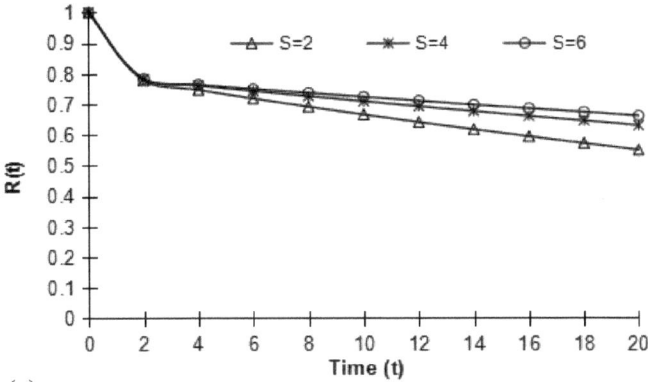

(a)

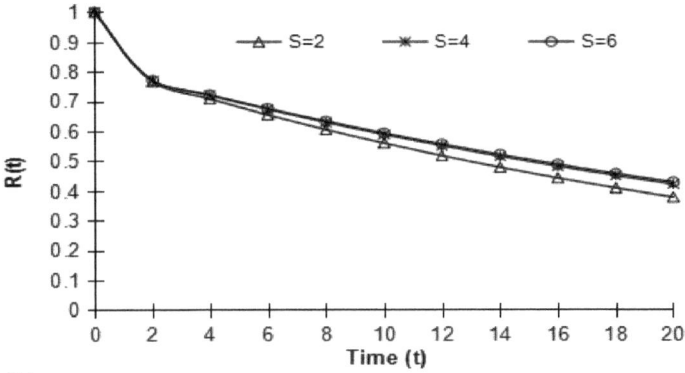

(b)

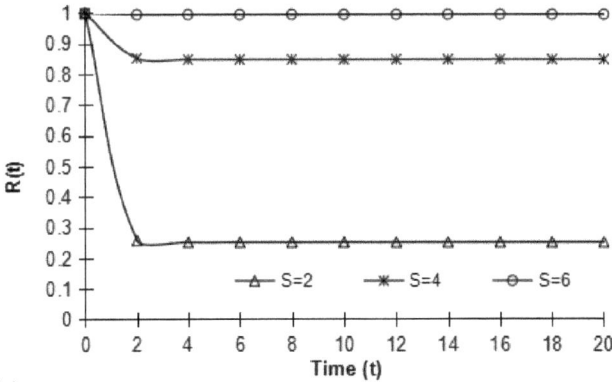

(c)

FIGURE 4.3 (a) Effect of number of standby machines on the reliability for $(a,b) = (0.5, 0.3)$; (b) effect of number of standby machines on the reliability for $(a,b) = (1, 1)$; (c) effect of number of standby machines on the reliability for $(a,b) = (1.5, 1.5)$.

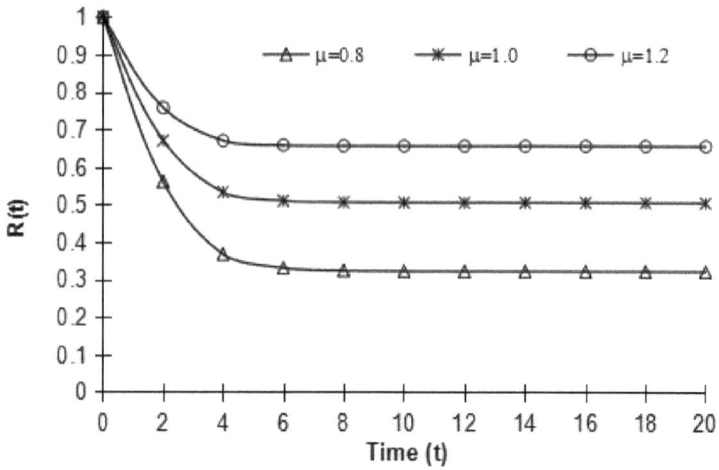

FIGURE 4.4   Effect of repair rate (μ) on the reliability for $(a,b)=(0.5,0.3)$.

Figures 4.4 and 4.5 are plotted to reveal the impact of repair rate (μ) on the reliability for $(a,b)=(0.5,0.3)$ and $(a,b)=(1.5,1.5)$, respectively. We see that the reliability increases with repair rate (μ); the reliability for $(a,b)=(0.5,0.3)$ has lower values in comparison to $(a,b)=(1.5,1.5)$. Figure 4.6 represents the reliability for $(a,b)=(0.5,0.3)$ for different values of the number of repairmen.

Figures 4.7 and 4.8 display the effects of the number of spares and repairmen, respectively, on the total cost function. The total cost exhibits the convexity with respect to both the number of spares and repairmen, and minimum cost is achieved for $S=6$ and $R=3$ as shown in Figures 4.7 and 4.8, respectively.

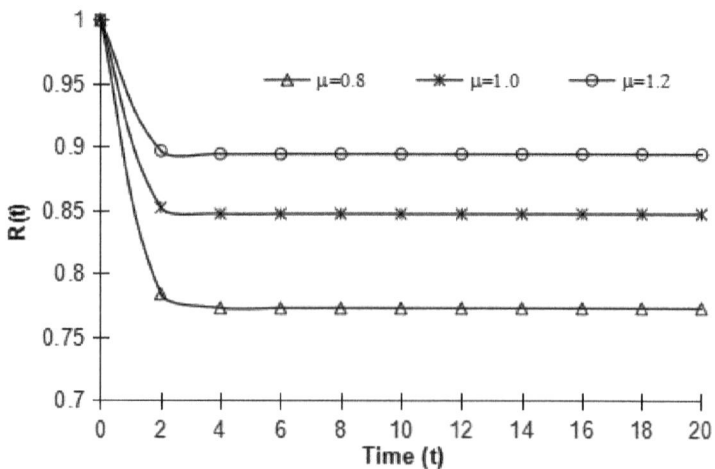

FIGURE 4.5   Effect of repair rate (μ) on the reliability for $(a,b)=(1,1)$.

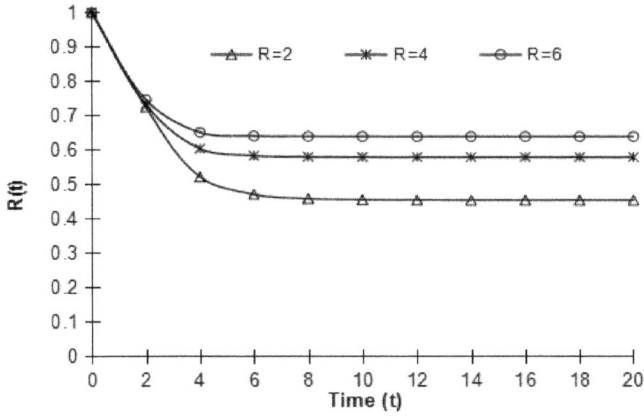

FIGURE 4.6 Effect of number of repairmen on the reliability for $(a, b) = (1.5, 1.5)$.

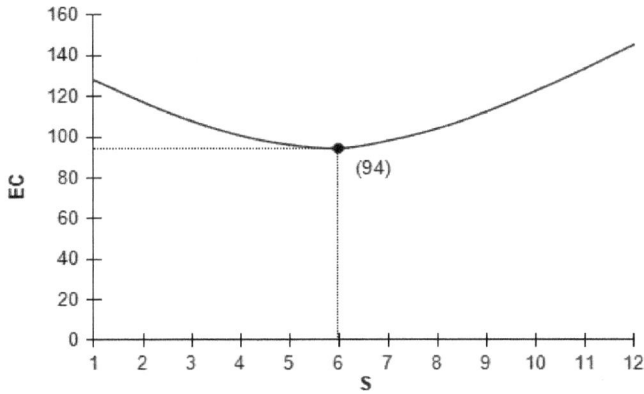

FIGURE 4.7 Expected cost by varying the number of standbys.

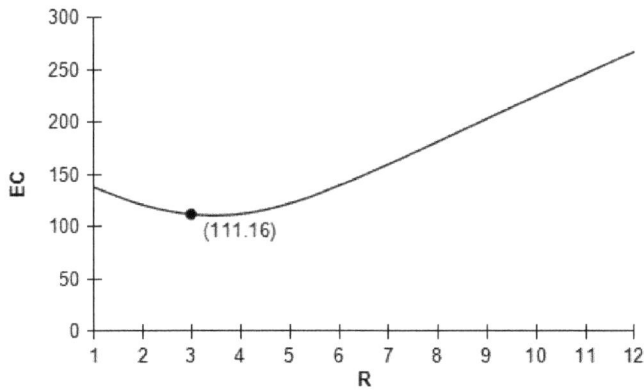

FIGURE 4.8 Expected cost by varying the number of repairmen.

4.6 CONCLUSION

In the concerned chapter, transient analysis of Markov model for fault tolerance machining system having state dependent rates as well as cold standby support has been facilitated. Some performance indices, namely, average counts of broken down machines, average counts of spares, average counts of busy/idle repairmen and operating utilization, etc. are provided. Our study will provide a valuable support for the performance prediction and estimation of inventory of spares and repairmen crew. The sensitivity analysis conducted demonstrates the usefulness of investigation done for the system engineers to find out the appropriate combination of repairmen and spares at optimum cost.

REFERENCES

Agnihothri, S. R. 1989. Inter-relationship between performance measures for the machine repairmen problem. *Naval Research Logistic*. 36:265–271.

AL-Seedy, R. O., 1992. The service Erlangian machine interference with balking. *Microelectronics Reliability*. 32:705–710.

Baker, K. N. 1973. A note on operating policies for the queue M/M/1 with exponential start-ups. *INFOR: Information Systems and Operational Research*.11:71–72.

Blackburn, J. D. 1972. Optimal control of a single server with balking and reneging. *Management Science*. 29:307–319.

Chiang, D. T. and Niu, S. C. 1981. Reliability of a consecutive K-out-of-N: G system. *IEEE Transaction on Reliability*. 30:87–90.

Chikara, D., Jain, M. and Baghel, K. P. S. 2006. Interdependent machine repair problem with controllable arrival rates and spares. *Acta Ciencia Indica Mathematica*. 32:569–574.

Dhakad, M. R. and Jain, M. 2016. Finite controllable Markovian model with balking and reneging. *International Journal of Science Technology and Engineering*. 2:36–45.

Gaver, D. P. and Lehoezky, J. P. 1977. A diffusion approximation solution for repairmen problem with two types of failures. *Management Science*. 24:71–81.

Goel, L. R., Gupta, R. and Singh, S. K. 1985. Cost analysis of a two units cold standby system with two types of operation and repair. *Microelectronics Reliability*. 25:71–75.

Gupta, S. M. 1995. Interrelationship between controlling arrival and service in queueing systems. *Computer and Operations Research*. 22:1005–1014.

Jain, M. 1997a. (m, M) machine repair problem with spares and state-dependent rates: A diffusion process approach. *Microelectronics Reliability*. 37:929–933.

Jain, M. 1997b. Optimal N-policy for single server Markovian queue with breakdown, repair and state dependent arrival rate. *International Journal of Management System*. 13:245–260.

Jain, M. 2013. Transient analysis of machining systems with service interruption, mixed standbys and priority. *International Journal of Mathematics in Operational Research*. 5:604–625.

Jain, M. and Baghel, K. P. S. 2001. A multi-component repairable system with spares and state-dependent rates. *Nepali Mathematical Science Report*. 19:81–92.

Jain, M. and Chauhan, D. 2010. Optimal control policy for state dependent queueing model with service interruption, setup and vacation. *Journal of Informatics and Mathematical Sciences*. 2:171–181.

Jain, M. and Gupta, R. 2018. N-policy for redundant repairable system with multiple types of warm standbys with switching failure and vacation. *International Journal of Mathematics in Operational Research*.13:419–449.

Jain, M. and Lata, P. 1994. M/M/R machine repair problem with reneging. *Journal of Engineering and Applied Sciences.* 13:139–143.

Jain, M. and Mittal, R. 2008. Transient analysis of channel allocation in cellular radio network with balking and reneging. *Journal of Computer Society of India.* 38:11–18.

Jain, M. Rakhee and Singh, M. 2004. Bi-level control of degraded machining system with warm standbys, setup and vacation. *Applied Mathematical Modeling.* 28:1015–1026.

Jain, M. and Sanga, S. S. 2019. Admission control for finite capacity queueing model with general retrial times and state dependent rates. *International Journal of Mathematics in Operational Research.* 1–44. doi: 10.3934/jimo.2019073.

Jain, M., Sharma, G. C., Baghel, K. P. S. and Shinde, V. 2004. Performance modeling of machining system with mixed standby components, balking and reneging. *International Journal of Engineering.* 17:169–180.

Jain, M., Sharma, G. C. and Rani, V. 2014. M/M/R+r machining system with reneging, spares and interdependent controlled rates. *International Journal of Mathematics in Operational Research.* 6:665–679.

Jain, M., Sharma, G. C. and Sharma, R. 2008. Performance modeling of state dependent system with mixed standbys and two modes of failure. *Applied Mathematical Modeling.* 32:712–724.

Jain, M., Sharma, G. C. and Sharma, V. 2009. Machine repair problem with two types of spares, set up and multiple vacations. *Ganita.* 60:111–125.

Jain, M., Sharma, R. and Meena, R. K. 2020. Performance modeling of fault tolerant machining system with working vacation and working breakdown, *Arabian Journal for Science Engineering.* 44:2825–2836.

Jain, M., Shekhar, C. and Shukla, S. 2016. A time-shared machine repair problem with mixed spares under N-policy. *Journal of Industrial Engineering International.* 12:145–157.

Jain, M., Singh, M. and Baghel, K. P. S. 2000. M/M/C/K/N machine repair problem with balking, reneging, spares and additional repairman. *GSR.* 26–27:49–60.

Kumar, K., Jain, M. and Shekhar, C. 2019. Machine repair system with F-policy, two unreliable servers and warm standbys. *Journal of Testing and Evaluation.* 47:1–23.

Maheshwari, S., Sharma, P. and Jain, M. 2010. Machine repair problem with K-type warm spares, multiple vacations for repairmen and reneging. *International Journal of Engineering Science and Technology.* 2:252–258.

Moustafa, M. S. 1998. Transient analysis of reliability with and without repair for K-out-of-N: G systems with M failure modes. *Reliability Engineering and System Safety.* 59:317–320.

Sharma, Y. K., Sharma, G. C. and Jain, M. 2017. Analysis of a queueing model with balking for buffer sharing in ATM. *Analysis.* 13:1–11.

Shinde, V., Sharma, G. C. and Jain, M. 2011. A parallel system sustain by standby units with failure in bulk. *Quality Control and Applied Statistics.* 56:153–154.

Singh, C. J. and Jain, M. 2007. (m,M) machine repair problem with spares and reneging. *Pakistan Journal of Statistics.* 23:23–35.

Singh, C. J., Jain, M. and Kumar, B. 2013. Analysis of unreliable bulk queue with state dependent arrival. *Journal of Industrial Engineering International.* 9:1–9.

Wang, K. H. and Ke, J. C. 2000. A recursive method to the optimal control of an M/G/1 queueing system with finite capacity and infinite capacity. *Applied Mathematical Modeling.* 24:899–914.

5 Performance Analysis of Markov Retrial Queueing Model under Admission Control F-Policy

Rakesh Kumar Meena
Institute of Science, Banaras Hindu University,
Varanasi, India

Pankaj Kumar
Indian Institute of Technology Roorkee, Uttarakhand, India

CONTENTS

5.1 INTRODUCTION

The formation of queues can be observed everywhere in real-life scenarios, including in front of the post office, schools, train ticket counters, shopping malls. For both customers and system organizers, the creation of long queues and service delays are the main issues. In some real-life situations, as well as in the service sectors (e.g., production and manufacturing, software and communications networks, etc.), clients (e.g., calls, employees, queries, notifications, broken equipment, etc.) arrive in a system for operation, but if the service is inaccessible, they are required to quit the service region and are guided to a simulated location called an orbit. Customers from orbit retry for the operation after a random period. Such reattempts within the service give rise to individual queues known *retrial queues*.

Significant research has been undertaken in the direction of retrial queues, which can be seen in the survey articles [Artalejo, 1999a and 1999b]. Wüchner et al. [2009] suggested that the homogenous multiserver finite source retrial queue be generalized where all performance metrics are checked using the MOSEL-2 tool. Artalejo and

65

Lopez-Herrero [2012] studied a finite population retrial queueing model using the block-structured state-dependent event (BSDE) approach to investigate the system state's limiting distribution and waiting time. Wang and Zhang [2013] analyzed the unobservable and observable retrial queue of the single server by examining the two situations with regard to specific information levels. Jain et al. [2015] set up a double-orbit, finite capacity retrial queue Markov model with an unreliable server to obtain both transient and steady-state probabilities, as well as different performance indices along with cost function.

Very few researchers have incorporated the working vacation (WV) concept to facilitate the queueing analysis of some realistic queueing circumstances. Due to the real-time applications of queueing framework with a WV from the server point of view, such an analysis has drawn the interest of several researchers in the queues during the last decade. Servi and Finn [2002] first proposed the idea of a WV where the server performs the service at a reduced pace, rather than stopping the service altogether over the vacation time. Wang et al. [2009] evaluated the Markovian machining system with a WV using MAPLE software to calculate the queue size distribution and many other system metrics. Jain and Jain [2010] addressed an unreliable Markovian queue with vacation and established various system indices using the geometric matrix approach. Yang and Wu [2015] analyzed an unreliable Markovian queueing problem operating under N-policy with WV and established matrix-form expressions for stationary probability distribution for mean queue length and other system metrics. Sethi et al. [2019] conducted the machining system's transient state analysis incorporating the feature of F-policy and WV. Recently, Jain et al. [2020] investigated the repairable machining system by combining modified server vacation principles along with imperfect recovery. The analytical results are obtained by using the recursive technique after including the supplementary variable. Furthermore, the quasi-Newton approach was used in this research work to determine the optimum parameters by reducing the overall cost.

To ensure the system's smooth operation and service standard, the customers' entry in the system should be monitored, and this can be achieved by introducing F-policy admission controls as stated in the queueing literature. Gupta [1995] first proposed the idea of F-policy. As per the F-policy, the arrival of customers into the service sector is halted when the system reached maximum capacity. Furthermore, as the queue size falls to a predefined threshold value "F", the customers are permitted to join the service sector. Nowadays, it is recognized that the monitoring of arrivals is the most significant issue in real-time scenarios in order to promote the level of service of waiting customers. To control entry of customers, Wang et al. [2007] examined the F-policy finite capacity queueing model. Kumar and Jain [2013] established the distribution of the queue size using a recursive approach for the Markovian queueing model functioning under both F-policy and N-policy. Jain et al. [2017] examined an MRP with a WV and F-policy. They obtained a steady-state queue size distribution by successive over-relaxation (SOR) approach in their analysis. Jain and Meena [2017a] analyzed a fault-tolerant system (FTS) with WV and controllable server and the performance metrics are obtained by using SOR approach in their investigation. Several queue theorists used the Runge-Kutta (R-K) approach to pursue the transient analysis of the queueing model [Jain and Meena, 2017b; Jain and Meena, 2018; Sethi and Bhagat, 2019].

In this chapter, the retrial queueing system is developed with the features of F-policy, WV and balking. The queue size distribution of considered system is computed using the R-K approach. The rest of the chapter is arranged as follows. In section 5.2, we give a description of the model and notations. The model governing equations are stated in section 5.3. In section 5.4, different metrics of the system are described in terms of probabilities of system state. The numerical results and system sensitivity analysis are given in section 5.5. Finally, section 5.6 draws the findings of the model under study by addressing the noble features and the potential direction.

5.2 MODEL DESCRIPTION

We formulate the Markovian retrial queueing model in this section for WV operating under admission control policy and start-up period. The F-policy's purpose is to monitor the entry of customers into the system. The underlying principles and notations used for model creation are described as:

The arriving customer follows a Poisson fashion to enter the system with a parameter λ. Due to an overcrowded system, the customers may be discouraged from joining the queue, and it is observed that the customers arriving for the service join the system and balk with probabilities p and $\bar{p} = 1 - p$, respectively. During a normal busy period the customers are served by a single server with mean $1/\mu$ according to an exponential distribution (exp. D). On entry, if the customer finds that the server is busy, he or she is required to wait for the service in the orbit and from orbit customer retries for the service according to an Exp. dist. with parameter γ. If the server is not occupied by the customer, the arriving customers get served immediately and leave the system. Once the system reaches its capacity M, the setup time is required to restrict the upcoming customers until the number of customers ceases to the predefined value F. The setup time is considered as follows Exp. dist. with rate ε. Furthermore, it is assumed that the server needs a start-up period before enabling customer entry through an Exp. dist. with parameter θ. The server moves to a WV with probability ξ until the device is empty, and it proceeds to do service at a slow pace; with complementary probability $1 - \xi$, the server stays idle. The service times during a WV period follows an Exp. dist. with mean $1/\mu_v$. The first-come-first-served discipline (FCFS) is being pursued to serving the customers.

We construct the model governing equations based on the Markov bi-variate process.

Let us define some notations as:

$\varsigma(\tau)$: Denotes the number of customers in the system at the time τ.
$\zeta(\tau)$: Denotes the status of the server at the time τ.

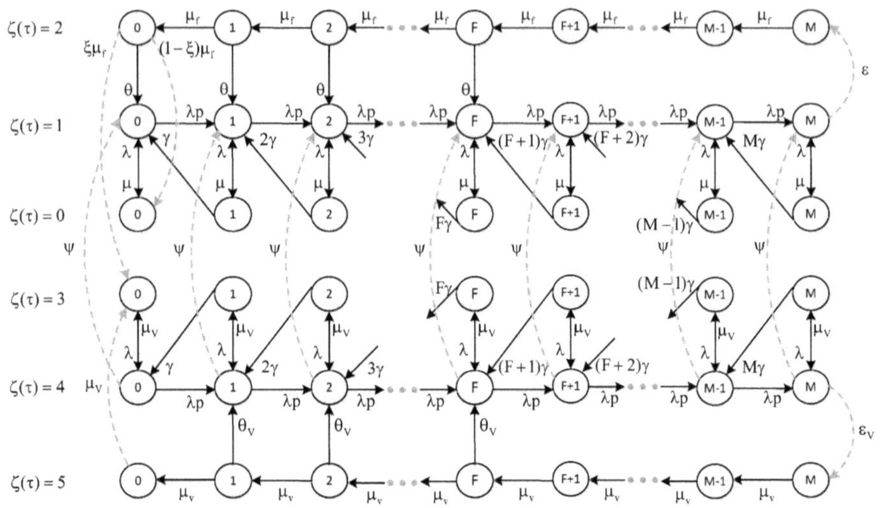

FIGURE 5.1   State transition diagram of M/M/1/WV queue with retrial orbit.

Now, we describe the server status (see Figure 5.1) as follows:

$$
\zeta(t) = \begin{cases}
0, & \text{Server is normally busy and customers are obliged to join the orbit,} \\
1, & \text{Server is normally busy, and customers will enter the system,} \\
2, & \text{Server is normally busy, and customers can not join the system,} \\
3, & \text{Server is busy in WV state and customers are obliged to join the orbit,} \\
4, & \text{Server is busy in WV state and customers are } \textit{allowed to enter the system,} \\
5, & \text{Server is busy in WV state and customers are not } \textit{allowed to enter the system.}
\end{cases}
$$

It is noted that $\{\varsigma(\tau), \zeta(\tau):\tau \geq 0\}$ is a continuous-time Markov process with state-space

$$
\Omega = \{(i,n)|i = 0,3; n = 0,1,...,M\} \cup \{(i,n)|i = 1,4; n = 0,1,...,M\}
$$
$$
\cup \{(i,n)|i = 2,5; n = 0,1,...,M\}.
$$

The transient state probabilities of the system states are defined as follows:

$Q_{i,n}(\tau)$: The probability that at time τ there are n ($1 \leq n \leq M$) failed machines in the system and the server is in state $\zeta(\tau) = i$; $0 \leq i \leq 5$.

5.3   TRANSIENT-STATE GOVERNING EQUATIONS

The Markov retrial queueing model's governing differential-difference equations framed based on the birth-death process are as follows:

1. The differential-difference equations for the normal retrial state have been framed by using the law of conservation of flows as follows:

$$Q'_{0,0}(\tau) = -\lambda Q_{0,0}(\tau) + \mu Q_{1,0}(\tau) + (1-\xi)\mu_f \, Q_{1,0}(\tau) \tag{5.1}$$

$$Q'_{0,n}(\tau) = -(n\gamma + \lambda)Q_{0,n}(\tau) + \mu Q_{1,n}(\tau); \ 1 \le n \le M-1 \tag{5.2}$$

$$Q'_{0,M}(\tau) = -(M\gamma + \lambda)Q_{0,M}(\tau) + \mu Q_{1,M}(\tau) \tag{5.3}$$

2. By equating out-flows from the normal busy state with the sum of in-flows from the other state to the normal busy states, the following equations are constructed:

$$Q'_{1,0}(\tau) = -(\mu + \lambda p)Q_{1,0}(\tau) + \lambda Q_{0,0}(\tau) + \gamma Q_{0,1}(\tau) + \theta Q_{2,0}(\tau) + \psi Q_{4,0}(\tau) \tag{5.4}$$

$$Q'_{1,n}(\tau) = -(\mu + \lambda p)Q_{1,n}(\tau) + \lambda p Q_{1,n-1}(\tau) + (n+1)\gamma Q_{0,n+1}(\tau) + \theta Q_{2,n}(\tau)$$
$$+ \lambda Q_{0,n}(\tau) + \psi Q_{4,n}(\tau); \ 1 \le n \le F \tag{5.5}$$

$$Q'_{1,n}(\tau) = -(\lambda p + \mu)Q_{1,n}(\tau) + \lambda p Q_{1,n-1}(\tau) + \lambda Q_{0,n}(\tau) + (n+1)\gamma Q_{0,n+1}(\tau)$$
$$+ \psi Q_{4,n}(\tau); \ F+1 \le n \le M-1 \tag{5.6}$$

$$Q'_{1,M}(\tau) = -(\varepsilon + \mu)Q_{1,M}(\tau) + \lambda p Q_{1,M-1}(\tau) + \lambda Q_{0,M}(\tau) + \psi Q_{4,M}(\tau) \tag{5.7}$$

3. To build the differential-difference equations for the normal F-policy, we apply the laws of flow conservation and obtain:

$$Q'_{2,0}(\tau) = -(\mu_f + \theta)Q_{2,0}(\tau) + \mu_f Q_{2,1}(\tau) \tag{5.8}$$

$$Q'_{2,n}(\tau) = -(\mu_f + \theta)Q_{2,n}(\tau) + \mu_f Q_{2,n+1}(\tau); \ 1 \le n \le F \tag{5.9}$$

$$Q'_{2,n}(\tau) = -\mu_f Q_{2,n}(\tau) + \mu_f Q_{2,n+1}(\tau); \ F+1 \le n \le M-1 \tag{5.10}$$

$$Q'_{2,M}(\tau) = -\mu_f Q_{2,M}(\tau) + \varepsilon Q_{1,M}(\tau) \tag{5.11}$$

4. The in-flows and out-flows are equated to frame the transient equations as:

$$Q'_{3,0}(\tau) = -\lambda_v Q_{3,0}(\tau) + \mu_v Q_{4,0}(\tau) + \mu_v Q_{5,0}(\tau) + \xi\mu_f Q_{2,0}(\tau) \tag{5.12}$$

$$Q'_{3,n}(\tau) = -(\lambda_v + n\gamma_v)Q_{3,n}(\tau) + \mu_v Q_{4,n}(\tau); \ 1 \le n \le M \tag{5.13}$$

5. For the WV states, the following equations are built by implementing flow conservation laws:

$$Q'_{4,0}(\tau) = -(\psi + \mu_v + \lambda p)Q_{4,0}(\tau) + \lambda Q_{3,0}(\tau) + \gamma Q_{3,1}(\tau) + \theta_v Q_{5,0}(\tau) \quad (5.14)$$

$$Q'_{4,n}(\tau) = -(\psi + \mu_v + \lambda p)Q_{4,n}(\tau) + \lambda Q_{3,n}(\tau) + (n+1)\gamma Q_{3,n+1}(\tau) \\ + \lambda p Q_{4,n-1}(\tau) + \theta_v Q_{5,n}(\tau); \ 1 \le n \le F \quad (5.15)$$

$$Q'_{4,n}(\tau) = -(\psi + \mu_v + \lambda p)Q_{4,n}(\tau) + \lambda Q_{3,n}(\tau) + (n+1)\gamma Q_{3,n+1}(\tau) \\ + \lambda_v p Q_{4,n}(\tau); \ F+1 \le n \le M-1 \quad (5.16)$$

$$Q'_{4,M}(\tau) = -(\psi + \mu_v + \varepsilon_v)Q_{4,M}(\tau) + \lambda p Q_{4,M}(\tau) + \lambda Q_{3,M}(\tau) \quad (5.17)$$

6. By equating the in-flows and out-flows equations for the WV F-policy level, we obtain:

$$Q'_{5,0}(\tau) = -(\theta_v + \mu_v)Q_{5,0}(\tau) + \mu_v Q_{5,1}(\tau) \quad (5.18)$$

$$Q'_{5,n}(\tau) = -(\mu_v + \theta_v)Q_{5,n}(\tau) + \mu_v Q_{5,n+1}(\tau); \ 1 \le n \le F \quad (5.19)$$

$$Q'_{5,n}(\tau) = -\mu_v Q_{5,n}(\tau) + \mu_v Q_{5,n+1}(\tau); \ F+1 \le n \le M-1 \quad (5.20)$$

$$Q'_{5,M}(\tau) = -\mu_v Q_{5,M}(\tau) + \varepsilon_v Q_{4,M}(\tau) \quad (5.21)$$

We inflict the initial condition as $Q(1) = 1$ and $Q(0) = 0$ for the determination of transient state probabilities by recognizing that the analytical approach to solving the system equation is quite monotonous. We use a numerical method based on the 4th order R-K to solve the equation system, Eqs. (5.1–5.21).

5.4   PERFORMANCE MEASURES

To examine the behavior of the system designed numerous system metrics are defined as follows in terms of the transient probabilities:

1. Expected number of customers in the system at a time τ is given as

$$L_S(\tau) = \sum_{i=0}^{5}\sum_{n=0}^{M} n\left(Q_{i,n}(\tau)\right) \quad (5.22)$$

2. The probability of the server being in F-policy is given as

$$Q_f(\tau) = \sum_{n=0}^{M} Q_{2,n}(\tau) + \sum_{n=0}^{M} Q_{5,n}(\tau) \tag{5.23}$$

3. The probability of the server being in normal busy state is given by

$$Q_B(\tau) = \sum_{n=0}^{M} Q_{1,n}(\tau) + \sum_{n=0}^{M} P_{2,n} \tag{5.24}$$

4. The probability of the server being in a WV state is given by

$$Q_W(\tau) = \sum_{n=0}^{M} Q_{4,n}(\tau) + \sum_{n=0}^{M} P_{5,n}(\tau) \tag{5.25}$$

5. The probability of the system being in retrial orbit is given as

$$Q_o(\tau) = \sum_{n=0}^{M} Q_{0,n}(\tau) + \sum_{n=0}^{M} Q_{3,n}(\tau) \tag{5.26}$$

6. Cost function

The cost elements per unit of the total cost are denoted by:

c_h: Holding price of every customer present in the system per unit period
c_b: Cost per unit time associated to the server is busy in a normal state
c_w: Cost per unit time associated to the server is in WV state
c_f: Cost per unit time associated with service rate during F-policy
c_v: Cost per unit time associated with service rate during WV
c_m: Cost per unit time associated with service rate

By considering the above costs associated with the different system metrics, we are now formulating the cost structure as follows:

$$C(\tau,\mu) = c_h L_S(\tau) + c_w Q_w(\tau) + c_b Q_B(\tau) + c_f \mu_f + c_v \mu_v + c_m \mu \tag{5.27}$$

5.5   NUMERICAL RESULTS

This segment analyzes the impact of various system parameters on specific system metrics such as $L_S(\tau)$, $Q_B(\tau)$, etc. In MATLAB software, we calculate the numerical results by encoding the computer program. The subroutine ode45 is used to solve the system of differential Eqs. (5.1–5.21), of different system states. We set default system parameters for illustration purposes as

$$M = 8,\ F = 2,\ \lambda = 0.3,\ \mu_f = 0.8, \psi = 0.8,\ \gamma = 0.5,\ \varepsilon = 0.2,\ \mu = 2.0,$$
$$\theta = 0.03,\ \gamma_V = 0.15,\ \mu_V = 0.75,\ \theta_V = 0.015,\ p = 0.6.$$

Cost set-I: $c_h = \$120,\ c_w = \$40,\ c_b = \$70,\ c_f = \$70,\ c_v = \$45,\ c_m = \$60$.
Cost set-II: $c_h = \$120,\ c_w = \$30,\ c_b = \$40,\ c_f = \$20,\ c_v = \$35,\ c_m = \$30$.
  The numerical findings are shown in Figures 5.2–5.7 and Tables 5.1–5.4 for conducting the sensitivity study. We have the following results, focused on numerical experiments:

1. **Effect of τ:** From Figures 5.2–5.7 and Tables 5.1–5.4, we see that the mean queue length, probability of busy state and the cost function increase as time goes up, but the probability of server is being in WV state decreases as time increase. It is clear that the impact of time (τ) diminished as time goes up, which shows that after a certain time, the system becomes stable as such there is no further change in $L_S(\tau)$.
2. **Effect of λ:** From Table 5.1, it is observed that as λ increases $L_S(\tau)$ and $Q_B(\tau)$ increases but $Q_w(\tau)$, $Q_f(\tau)$ and $Q_o(\tau)$, decreases. From Figure 5.3, we see that the mean queue length $L_S(\tau)$ is increasing as λ increases; the effect of λ is not much remarkable after a certain time period, *i.e.*,$\tau = 20$.
3. **Effect of μ:** In Table 5.2, as service rate (μ) increases, $L_S(\tau)$ and $Q_B(\tau)$, probability of customer in retrial orbit $Q_o(\tau)$ decreases with μ. Further, the probability that the server is being busy in WV $Q_w(\tau)$ and F-policy state

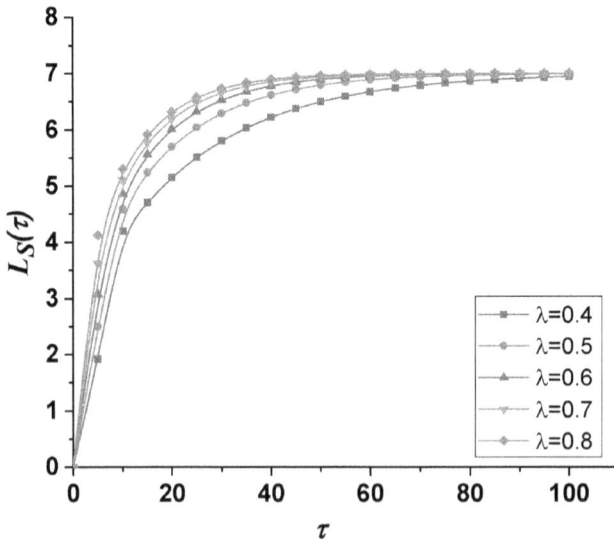

FIGURE 5.2 $L_S(\tau)$ vs. τ for varying values of λ.

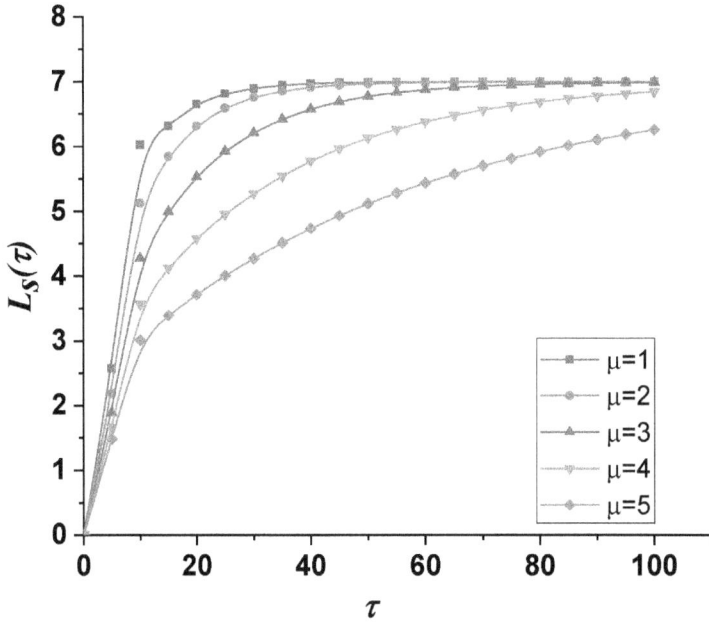

FIGURE 5.3 $L_S(\tau)$ vs. τ for varying values of μ.

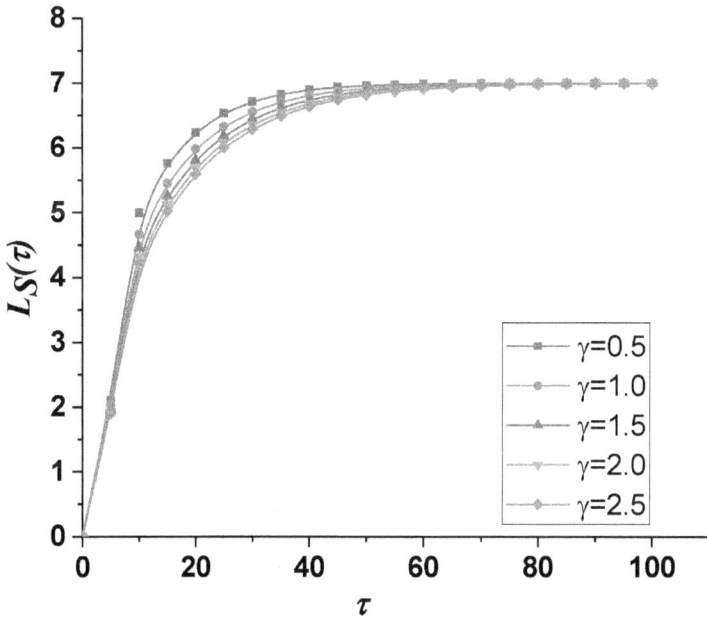

FIGURE 5.4 $L_S(\tau)$ vs. τ for varying values of γ.

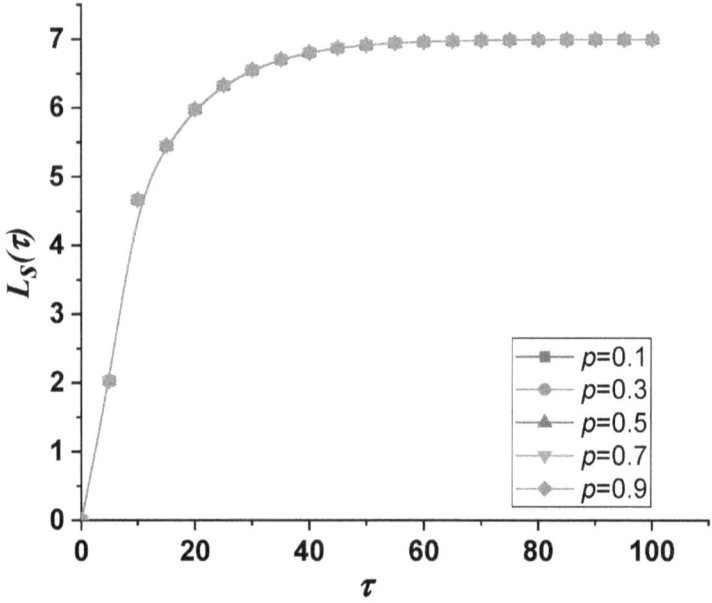

FIGURE 5.5 $L_S(\tau)$ vs. τ for varying values of p.

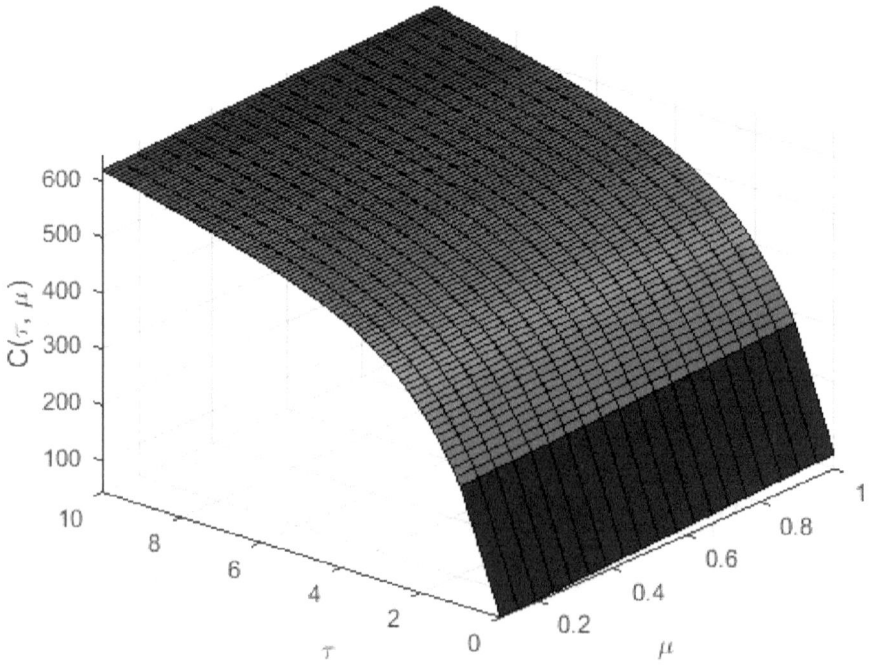

FIGURE 5.6 $C(\tau,\mu)$ vs. time τ and μ.

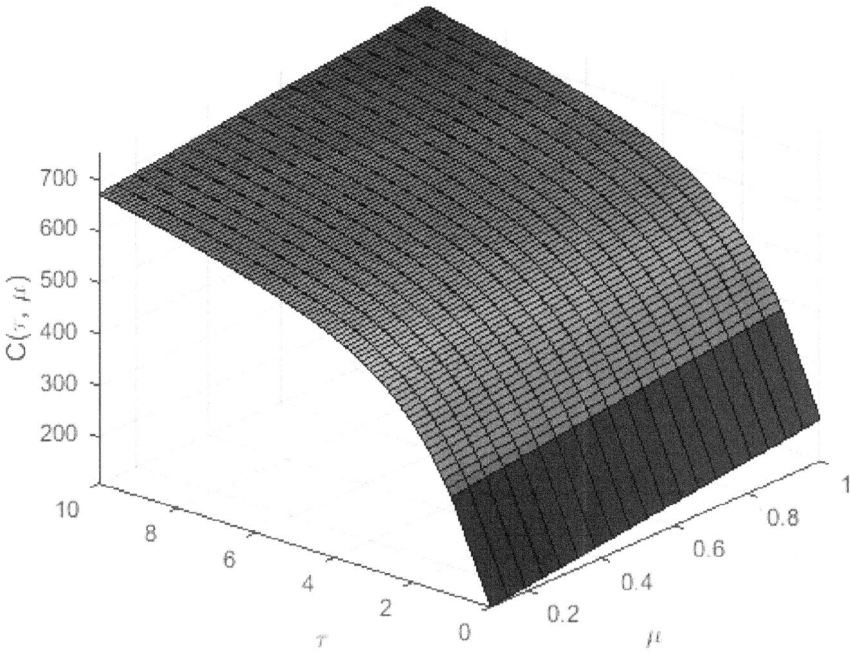

FIGURE 5.7 $C(\tau,\mu)$ vs. time τ and $C(\tau,\mu)$.

$Q_f(\tau)$ increases with μ. From Figure 5.4, it can be seen that the expected number of customers in the system decreases with μ which is as per our expectation.
4. **Effect of θ:** In Table 5.3, we observe that as θ increases, $L_S(\tau)$ and $Q_B(\tau)$ increases. On the contrary, the probability of a server is in WV $Q_w(\tau)$, retrial orbit $Q_o(\tau)$ and F-policy state $Q_f(\tau)$ decreases.

TABLE 5.1
Performance Measures for the Varying Value of λ

λ	τ	$L_S(\tau)$	$Q_B(\tau)$	$Q_w(\tau)$	$Q_f(\tau)$	$Q_o(\tau)$
0.3	05	1.891395	0.631113	2.48E-09	0.368211	6.75E-04
	10	4.27391	0.703563	0.003883	0.249797	0.042757
	15	5.001269	0.772667	0.011221	0.182177	0.033934
0.5	5	2.922908	0.702569	4.91E-08	0.289874	7.56E-03
	10	5.822947	0.810853	0.018364	0.087204	0.083578
	15	6.530426	0.903275	0.029593	0.035508	0.031623
0.7	5	3.925209	0.741718	3.27E-07	0.228371	2.99E-02
	10	6.348627	0.866441	0.030456	0.037471	0.065633
	15	6.830805	0.929732	0.042479	0.009517	0.018271

TABLE 5.2

Performance Measures for the Varying Value of μ

μ	τ	$L_S(\tau)$	$Q_B(\tau)$	$Q_w(\tau)$	$Q_f(\tau)$	$Q_o(\tau)$
3	05	1.891395	0.631113	2.48E-09	0.368211	6.75E-04
	10	4.27391	0.703563	0.003883	0.249797	0.042757
	15	5.001269	0.772667	0.011221	0.182177	0.033934
5	5	1.47974	0.500622	2.75E-08	0.499126	2.52E-04
	10	3.010376	0.555224	0.002141	0.433413	0.009222
	15	3.391266	0.591191	0.005429	0.394288	0.009092
7	5	1.208703	0.412773	7.93E-08	0.587125	1.02E-04
	10	2.260192	0.446997	0.000836	0.549917	0.00225
	15	2.420881	0.462023	0.001984	0.533666	0.002327

TABLE 5.3

Performance Measures for the Varying Value of θ

θ	τ	$L_S(\tau)$	$Q_B(\tau)$	$Q_w(\tau)$	$Q_f(\tau)$	$Q_o(\tau)$
1	05	1.891395	0.631113	2.48E-09	0.368211	6.75E-04
	10	4.273911	0.703563	0.003883	0.249797	0.042757
	15	5.001383	0.772673	0.011237	0.182181	0.033908
2	5	1.891395	0.631114	1.97E-09	0.368212	6.75E-04
	10	4.298082	0.709813	0.001751	0.250494	0.037942
	15	5.042577	0.783836	0.00472	0.181528	0.029916
3	5	1.891396	0.631114	1.59E-09	0.368212	6.74E-04
	10	4.310819	0.712717	0.000897	0.250874	0.035511
	15	5.063381	0.788548	0.002311	0.181208	0.027933

TABLE 5.4

Performance Measures for the Varying Value of γ

γ	τ	$L_S(\tau)$	$Q_B(\tau)$	$Q_w(\tau)$	$Q_f(\tau)$	$Q_o(\tau)$
1	05	1.891395	0.631113	2.48E-09	0.368211	6.75E-04
	10	4.273911	0.703563	0.003883	0.249797	0.042757
	15	5.001383	0.772673	0.011237	0.182181	0.033908
3	5	1.708253	0.707814	2.48E-09	0.291529	6.57E-04
	10	3.603521	0.783898	0.003389	0.180347	0.032367
	15	4.285772	0.819604	0.009094	0.143881	0.027421
5	5	1.616971	0.742618	2.48E-09	0.256734	6.48E-04
	10	3.352484	0.815069	0.00323	0.152391	0.029309
	15	4.013714	0.841728	0.008412	0.12457	0.02529

5. **Effect of** γ: From Table 5.4, we see that as γ increases than expected number of customers in the system, the probability of server is being a busy state, shows significant changes as γ increases.
6. **Effect of** p: As the joining probability (p) of the customer's increases, then increases as time grows. It is noticed from Figure 5.5 that as p increases, the expected number of customers in the system, probability of server is being busy state, remain almost constant, i.e., p does not reveal much impact on various indices.

5.6 CONCLUSIONS

We investigated the single-server retrial queueing model by introducing practical aspects such as balking, admission control policy and WV. We have established various performance indices, including mean system size, mean queue length, various system sate probabilities, etc., which are further used to construct the cost function. The applicability of F-policy, along with the WV and balking feature, can be understood in many real-time applications where the optimal F-policy would have improved control of customer admission. The present work may further be extended by considering general service time.

REFERENCES

Artalejo, J.R. 1999a. A classified bibliography of research on retrial queues. *Top*. 7: 187–211.
Artalejo, J.R. 1999b. Accessible bibliography on retrial queues. *Mathematical and Computer Modeling*. 30: 1–6.
Artalejo, J.R. and Lopez-Herrero, M.J. 2012. The single server retrial queue with finite population: a BSDE approach. *Operational Research*. 12: 109–131.
Gupta, S.M. 1995. Interrelationship between controlling arrival and service in queueing systems. *Computers and Operations Research*. 22: 1005–1014.
Jain, M., Bhagat, A. and Shekhar, C. 2015. Double orbit finite retrial queues with priority customers and service interruptions. *Applied Mathematics and Computation*. 253: 324–344.
Jain, M. and Jain, A. 2010. Working vacations queueing model with multiple types of server breakdowns. *Applied Mathematical Modeling*. 34: 1–13.
Jain, M. and Meena, R.K. 2017a. Markovian analysis of unreliable multi-components redundant fault tolerant system with working vacation and F-policy. *Cogent Mathematics*. 4: 1–17.
Jain, M. and Meena, R.K. 2017b. Fault tolerant system with imperfect coverage, reboot and server vacation. *Journal of Industrial Engineering International*. 13: 171–180.
Jain, M. and Meena, R.K. 2018. Vacation model for Markov machine repair problem with two heterogeneous unreliable servers and threshold recovery. *Journal of Industrial Engineering International*. 14: 143–152.
Jain, M., Meena, R.K. and Kumar, P. 2020. Maintainability of redundant machining system with vacation, imperfect recovery and reboot delay. *Arabian Journal for Science and Engineering*. 45: 2145–2161.
Jain, M., Shekhar, C. and Meena, R.K. 2017. Admission control policy of maintenance for unreliable server machining system with working vacation. *Arabian Journal for Science and Engineering*. 42: 2993–3005.

Kumar, K. and Jain, M. 2013. Threshold F-policy and N-policy for multi-component machining system with warm standbys. *Journal of Industrial Engineering International.* 9: 9–28.

Servi, L.D. and Finn, S.G. 2002. M/M/1 queues with working vacations (M/M/1/WV). *Performance Evaluation.* 50: 41–52.

Sethi, R. and Bhagat, A. 2019. Performance analysis of machine repair problem with working vacation and service interruptions. *AIP Conference Proceedings.* 2061: 1–8.

Sethi, R., Bhagat, A. and Garg, D. 2019. ANFIS based machine repair model with control policies and working vacation. *International Journal Mathematical Engineering and Management Sciences.* 4: 1522–1533.

Wang, J. and Zhang, F. 2013. Strategic joining in M/M/1 retrial queues. *European Journal of Operational Research.* 230: 76–87.

Wang, K.H., Chen, W.L. and Yang, D.Y. 2009. Optimal management of the machine repair problem with working vacation: Newton's method. *Journal of Computational and Applied Mathematics.* 233: 449–458.

Wang, K.H., Kuo, C.C. and Pearn, W.L. 2007. Optimal control of an M/G/1/K queueing system with combined F policy and start-up time. *Journal of Optimization Theory and Application.* 135: 285–299.

Wüchner, P., Sztrik, J. and de, M.H. 2009. Finite-source retrial queue with search for balking and impatient customers from the orbit. *Computer Networks* 53: 1264–1273.

Yang, D.Y. and Wu, C.H. 2015. Cost-minimization analysis of a working vacation queue with N-policy and server breakdowns. *Computer and Industrial Engineering.* 82: 151–158.

6 A Study on *Geo/G/1* Retrial Queueing System with Starting Failure and Customer Feedback

S. Pavai Madheswari, P. Suganthi
R.M.K. Engineering College, Chennai, India

K. Vasanthamani
Kalaignarkarunanidhi Institute of Technology,
Coimbatore, India

CONTENTS

6.1   INTRODUCTION

In the past two decades, significant contributions have been made to the analysis of queueing models with recurrent customers. Readers may refer Artalejo (1999), Gomez-Corral (2006) and Artalejo (2010) for the bibliography on retrial queues. The researchers in the field of retrial queues were mainly focused on the continuous time models. But, as a result of technological growth in the computer and telecommunication systems, at the end of twentieth century researchers started to analyze retrial queues in discrete time. Bruneel and Kim (1993), Takagi (1993) and Woodward (1994) discuss the discrete time queues in detail. It is observed that the discrete time queues are more appropriate to analyze the systems that have digital basic units such as bits, packets, slots and machine cycle time. The discrete time queues have been discussed by several researchers initiated by Meisling (1958) and further studied by

Atencia and Moreno (2004a, 2006), Artalejo and Gomez-Corral (2008) and Wu et al. (2013). Nobel (2016) presented a short survey on retrial queueing models in discrete time. Atencia (2017) has studied a Geo/G/1 retrial queueing system with priority services. Upadhyaya (2018) has done a performance analysis of a Geo/G/1 retrial queue under J-vacation policy. A single-server retrial queue with server vacation was studied by Gao and Wang (2019) wherein they considered two waiting buffers to model ATM networks. Recently, Lan and Tang (2019) have analyzed the performance of a "Discrete time Geo/G/1 retrial queue with non–pre-emptive priority, working vacations and vacation interruption". Kulshrestha and Shruti l (2020) analyzed a discrete time to model communication networks with second optional service and negative user arrivals. Lan and Tang (2019) have dealt with a discrete time Geo/G/1 retrial queueing system with probabilistic pre-emptive priority and balking customers, in which the server is subject to starting failures and replacements in the repair times may occur with some probability.

Server breakdown is an inevitable phenomenon in any service facility. But most of the papers in the queueing literature consider the queueing models where the servers are never subjected to failure, which is practically unrealistic. Actually, queueing systems with servers subjected to starting failures are a natural and important idea for modeling server unavailability due to external causes. Wang and Zhao (2007) studied a discrete time queue where the server may not be started successfully every time when the customer approaches for service. Customer impatience is one of the important features of any real time queueing situation, as customers do not like to spend more time on just waiting to be served – i.e., they may not join the system if they expect a long wait. In view of the importance of this balking phenomenon in the communication systems, many researchers are now paying attention to the queueing models with customer balking (Aboul-Hassan et al., 2005, 2008; Wei et al., 2015). Recently, Madheswari et al. (2019) have discussed an M/G/1 retrial queue in which they have taken customer balking as one of the constraints. In our present study we take into account the customer's dilemma whether to join the system or balk. The erroneous or lost messages are retransmitted in any communication system, and if a customer is not satisfied with the service received then the server may repeat the call in any call center facility. These situations are well studied by modelling them as a queueing system with customer feedback. Introduced by Takacs (1963), the feedback concept has been studied in continuous time (Choi and Kim, 1997; Krishna Kumar et al., 2002; Madan and Ai-Rawwash, 2005, etc.,) and also discussed for discrete time queueing systems (Atencia and Moreno, 2004b; Atencia et al., 2009).

The practical motivation for the model considered is the wireless sensor networks (WSN) deployed to monitor the forest fire or in the military surveillance that operates in discrete time environment. The WSN has a number of sensor nodes and a sink node. The sensor nodes communicate the data that it detected to the sink node through wireless links for further processing. The sensor nodes act as the server, and there is a possibility of failure while starting the transmission of the sensed data. The data packets not delivered to the sink node will be retransmitted. When a data packet arrives at a sensor node for transmission but the node is busy or down, the data may be lost. In this scenario of WSN the data to be transmitted, the sensor node, undelivered packets retransmitted, the data finding the sensor node busy is trying again and a few data are lost when the node is busy or down correspond to customer, server, feedback, retrial and balking in queueing terminology, respectively.

In this chapter, we analyze the retrial queueing system with customer feedback and balking where there is a possibility of server failure while starting the service in discrete time. We consider an early arrival system (EAS) where the departure has precedence over the arrivals. We expect this work to be the discrete time counterpart of the study on "M/G/1 retrial queueing system with customer balking and feedback where the server is prone to starting failure". The remainder of the chapter is organized as: Our model is described in detail in section 6.2. The Markov chain of our model is analyzed in section 6.3. Marginal generating functions are derived of the orbit size for different server states. The decomposition property is established in section 6.4 for the system size distribution. In section 6.5, we give graphical illustrations to show the impact of chosen parameters on some quantities of interest. Section 6.7 provides the concluding remarks.

6.2   OUR MODEL

We examine a discrete time single-server queueing system with recurrent customers in which the time line is divided into small subintervals denoted by $0, 1, 2 \ldots$. Arrivals, departures, retrials, server failures and repair completion are assumed to occur only near the end of the subintervals. Further, all queueing activities that make the server become available (departures and repair completion) occur in (t^-, t), and activities that make the server not available to serve the customers (arrivals, retrials, server failure) occur in (t, t^+) in the same order. These are the characteristics of the "early arrival system EAS". Primary arrivals from outside follow a geometrical distribution with probability a. If the server is not available for providing service, the customer chooses to join the orbit with probability $\bar{b}$ or to balk (leave the system once for all) with probability b. On the other hand, if the server is available upon arrival of a customer, his service begins at once. The customers waiting in the orbit are permitted to try again for their service as per the FCFS discipline. It is assumed that the inter retrial times follow an arbitrary distribution $\{r_h\}_{h=0}^{\infty}$. The generating function is $L(x) = \displaystyle\sum_{h=0}^{\infty} r_h x^h$. If the server is idle when a new or retrial customer arrives, he should "turn on" the sever to receive his service. With probability θ the server is turned on and starts serving the customer immediately. On the other hand, the server fails to start with probability $\bar{\theta} = 1 - \theta$ and the server undergoes repair. In this case, the customer who attempted to start the server has to join the orbit. The service times are assumed to be independent and identically distributed (i.i.d) random variables following arbitrary distributions $\{b_{1,j}\}_{j=1}^{\infty}$, generating functions $B_1(x) = \displaystyle\sum_{j=1}^{\infty} b_{1,j} x^j$ and n^{th} factorial moments $\beta_{1,n}$. The customer decides to leave the system after service completion with probability $\bar{v}$ or to join the orbit for another service with probability v $(\bar{v} + v = 1)$. The repair times are also i.i.d random variables and assumed to follow general distribution $\{b_{2,j}\}_{j=1}^{\infty}$, generating function $B_2(x) = \displaystyle\sum_{j=1}^{\infty} b_{2,j} x^j$ and n^{th} factorial moments $\beta_{2,n}$. We suppose $0 < a < 1$ and $0 < \theta < 1$ to avoid known cases.

Throughout the chapter, we will denote by $\bar{a} = 1 - a$. The load of the system is given by $\rho_1 + \rho_2$ where $\rho_1 = a\beta_{1,1}; \rho_2 = a\beta_{2,1}$.

6.3   THE MARKOV CHAIN

The model proposed in the previous section is described by discrete time Markov process $Y_i = (C_i, \zeta_{0,i}, \zeta_{1,i}, \zeta_{2,i}, N_i)$ at time i^+ where

$$C_i = \begin{cases} 0, & \text{if the server is free} \\ 1, & \text{if the server is busy} \\ 2, & \text{the server is under repair} \end{cases}$$

and N_i denotes the number of retrial customers. If $C_i = 0$, then $\zeta_{0,i}$ denotes the remaining retrial time, when $C_i = 1$, then $\zeta_{1,i}$ denotes the remaining service time and if $C_i = 2$, then $\zeta_{2,i}$ corresponds to the time required for completion of repair. The process $\{Y_i, i \in N\}$ can be shown as the Markov chain of the model under study. $\{(0,0);(0,h,g):h \geq 1, g \geq 1;(1,h,g):h \geq 1, g \geq 0;(2,h,g):h \geq 1, g \geq 1\}$.

It is required to derive the following steady-state distribution:

$$P(0) = \lim_{i \to \infty} P[C_i = 0, N_i = 0],$$

$$P(h,g) = \lim_{i \to \infty} P[C_i = 0, \zeta_{0,i} = h, N_i = g], h \geq 1, g \geq 1,$$

$$Q(h,g) = \lim_{i \to \infty} P[C_i = 1, \zeta_{1,i} = h, N_i = g], h \geq 1, g \geq 0,$$

$$R(h,g) = \lim_{i \to \infty} P[C_i = 2, \zeta_{2,i} = h, N_i = g], h \geq 1, g \geq 1,$$

of the Markov chain $Y_i = (C_i, \zeta_{0,i}, \zeta_{1,i}, \zeta_{2,i}, N_i)$.

The system of Kolmogorov equations for the stationary distribution is obtained as

$$P(0) = \bar{a}P(0) + \bar{a}vP(1,0), \tag{6.1}$$

$$P(h,g) = \bar{a}P(h+1,g) + \bar{p}r_h \bar{v}Q(1,g) + \bar{a}r_h vQ(1,g-1) + \bar{a}r_h R(1,g), h \geq 1, g \geq 1, \tag{6.2}$$

$$Q(h,g) = \delta_{0g}a\theta b_{1,h}P(0) + \bar{a}\theta b_{1,h}P(1,g+1) + (1-\delta_{0g})a\theta b_{1,h}\sum_{j=1}^{\infty}P(j,g)$$

$$+ (1-\delta_{0g})a\theta b_{1,h}vQ(1,g-1)$$

$$+ (a\theta b_{1,h}\bar{v} + v\bar{a}r_0\theta b_{1,h})Q(1,g) + \bar{a}r_0\theta b_{1,h}\bar{v}Q(1,g+1)$$

$$+ (1-\delta_{0g})ab\bar{Q}(h+1,g-1) + (\bar{a}+ab)Q(h+1,g)$$

$$+ (1-\delta_{0g})a\theta b_{1,h}R(1,g) + \bar{a}r_0\theta b_{1,h}R(1,g+1), h \geq 1, g \geq 0, \tag{6.3}$$

$$R(h,g) = \delta_{1g}a\bar{\theta}b_{2,h}P(0) + (1-\delta_{1g})a\bar{\theta}b_{2,h}\sum_{j=1}^{\infty}P(j,g-1) + \bar{a}\bar{\theta}b_{2,h}P(1,g)$$

$$+ (a\bar{\theta}b_{2,h}\bar{v} + \bar{a}r_0\bar{\theta}b_{2,h}v)Q(1,g-1)$$

$$+\bar{a}r_0\bar{\theta}\bar{v}b_{2,h}Q(1,g) + (1-\delta_{1g})a\bar{\theta}b_{2,h}vQ(1,g-2)$$
$$+(1-\delta_{1g})a\bar{\theta}b_{2,h}R(1,g-1) + \bar{a}r_0\bar{\theta}b_{2,h}R(1,g)$$

$$+(1-\delta_{1g})a\bar{b}R(h+1,g-1) + (ab+\bar{a})R(h+1,g), h \geq 1, g \geq 1, \quad (6.4)$$

where normalization condition is

$$P(0) + \sum_{h=1}^{\infty}\sum_{g=1}^{\infty}P(h,g) + \sum_{h=1}^{\infty}\sum_{k=0}^{\infty}Q(h,g) + \sum_{h=1}^{\infty}\sum_{g=1}^{\infty}R(h,g) = 1.$$

For solving the Eqs. (6.1)–(6.4), the PGFs are defined as:

$$\varphi_0(x,z) = \sum_{i=1}^{\infty}\sum_{g=1}^{\infty}P(h,g)x^h z^g; \quad \varphi_1(x,z) = \sum_{i=1}^{\infty}\sum_{g=0}^{\infty}Q(h,g)x^h z^g \quad \text{and}$$

$$\varphi_2(x,z) = \sum_{i=1}^{\infty}\sum_{g=1}^{\infty}R(h,g)x^h z^g$$

and the auxiliary generating functions as:

$$\varphi_{0,h}(z) = \sum_{g=1}^{\infty}P(h,g)z^g; \quad \varphi_{1,h}(z) = \sum_{g=0}^{\infty}Q(h,g)z^g \quad \text{and}$$

$$\varphi_{2,h}(z) = \sum_{g=1}^{\infty}R(h,g)z^g, h \geq 1.$$

Multiplying Eqs. (6.2)–(6.4) by z^g and taking summation over g, we have

$$\varphi_{0,h}(z) = \bar{a}\varphi_{0,h+1}(z) + \bar{a}r_h[\bar{v} + vz]\varphi_{1,1}(z) - \bar{a}r_h Q(1,0) + \bar{a}r_h\varphi_{2,1}(z), h \geq 1, \quad (6.5)$$

$$\varphi_{1,h}(z) = a\bar{\theta}b_{1,h}P(0) + \frac{\bar{a}}{z}\bar{\theta}b_{1,h}\varphi_{0,1}(z) + a\bar{\theta}b_{1,h}\varphi_0(1,z) + (\bar{v}+vz)(a+\frac{\bar{a}r_0}{z})\bar{\theta}b_{1,h}\varphi_{1,1}(z)$$

$$-\frac{\bar{a}r_0\bar{v}}{z}\bar{\theta}b_{1,h}Q(1,0) + (a\bar{b}z + ab + \bar{a})\varphi_{1,h+1}(z) + (a+\frac{\bar{a}r_0}{z})\bar{\theta}b_{1,h}\varphi_{2,1}(z), h \geq 1, \quad (6.6)$$

$$\varphi_{2,h}(z) = az\bar{\theta}b_{2,h}P(0) + az\bar{\theta}b_{2,h}\varphi_0(1,z) + \bar{a}\bar{\theta}b_{2,h}\varphi_{0,1}(z) + (az+\bar{a}r_0)(\bar{v}+vz)\bar{\theta}b_{2,h}\varphi_{1,1}(z)$$

$$-\bar{a}r_0\bar{v}\bar{\theta}b_{2,h}Q(1,0) + (az+\bar{a}r_0)\bar{\theta}b_{2,h}\varphi_{2,1}(z) + (a\bar{b}z+ab+\bar{a})\varphi_{2,h+1}(z), h \geq 1. \quad (6.7)$$

Substituting Eq. (6.1) in Eqs. (6.5)–(6.7), we get

$$\varphi_{0,h}(z) = \overline{a}\varphi_{0,h+1}(z) + \overline{a}r_h(\overline{v} + vz)\varphi_{1,1}(z) + \overline{a}r_h\varphi_{2,1}(z) - ar_hP(0), h \geq 1, \tag{6.8}$$

$$\varphi_{1,h}(z) = \frac{\overline{a}}{z}\theta b_{1,h}\varphi_{0,1}(z) + a\theta b_{1,h}\varphi_0(1,z) + (\overline{v} + vz)(a + \frac{\overline{a}r_0}{z})\theta b_{1,h}\varphi_{1,1}(z)$$

$$+ (a\overline{b}z + ab + \overline{a})\varphi_{1,h+1}(z) + (a + \frac{\overline{a}r_0}{z})\theta b_{1,h}\varphi_{2,1}(z) + \frac{z - r_0}{z}a\theta b_{1,h}P(0), h \geq 1, \tag{6.9}$$

$$\varphi_{2,h}(z) = az\overline{\theta}b_{2,h}\varphi_0(1,z) + \overline{a}\overline{\theta}b_{2,h}\varphi_{0,1}(z) + (\overline{v} + vz)(az + \overline{a}r_0)\overline{\theta}b_{2,h}\varphi_{1,1}(z)$$

$$+ (az + \overline{a}r_0)\overline{\theta}b_{2,h}\varphi_{2,1}(z) + (a\overline{b}z + ab + \overline{a})\varphi_{2,h+1}(z) + (z - r_0)a\overline{\theta}b_{2,h}P(0), h \geq 1. \tag{6.10}$$

Multiplying Eqs. (6.8)–(6.10) by x^h and taking summation over h, we obtain

$$\frac{x - \overline{a}}{x}\varphi_0(x,z) = \overline{a}(\overline{v} + vz)[L(x) - r_0]\varphi_{1,1}(z) - \overline{a}\varphi_{0,1}(z) + \overline{a}[L(x)$$
$$- r_0]\varphi_{2,1}(z) - a[L(x) - r_0]P(0), \tag{6.11}$$

$$\frac{x - (a\overline{b}z + ab + \overline{a})}{x}\varphi_1(x,z) = \frac{\overline{a}}{z}\theta B_1(x)\varphi_{0,1}(z) + a\theta B_1(x)\varphi_0(1,z)$$

$$+ \left[\left(\frac{az + \overline{a}r_0}{z}\right)(\overline{v} + vz)\theta B_1(x) - (a\overline{b}z + ab + \overline{a})\right]\varphi_{1,1}(z)$$

$$+ \left(\frac{az + \overline{a}r_0}{z}\right)\theta B_1(x)\varphi_{2,1}(z) + \left(\frac{z - r_0}{z}\right)a\theta B_1(x)P(0), \tag{6.12}$$

$$\frac{x - (a\overline{b}z + ab + \overline{a})}{x}\varphi_2(x,z) = az\overline{\theta}B_2(x)\varphi_0(1,z) + \overline{a}\overline{\theta}B_2(x)\varphi_{0,1}(z)$$
$$+ (az + \overline{a}r_0)(\overline{v} + vz)\overline{\theta}B_2(x)\varphi_{1,1}(z)$$

$$+ [(az + \overline{a}r_0)\overline{\theta}B_2(x) - (a\overline{b}z + ab + \overline{a})]\varphi_{2,1}(z) + (z - r_0)a\overline{\theta}B_2(x)P(0). \tag{6.13}$$

Choosing $x = 1$ in Eq. (6.11), it reduces to

$$a\varphi_0(1,z) = \overline{a}(\overline{v} + vz)[1 - r_0]\varphi_{1,1}(z) - \overline{a}\varphi_{0,1}(z) + \overline{a}[1 - r_0]\varphi_{2,1}(z) - a[1 - r_0]P(0).$$

Now using the above equation in Eqs. (6.12) and (6.13), we obtain

$$\frac{x-(a\bar{b}z+ab+\bar{a})}{x}\varphi_1(x,z)=\left(\frac{1-z}{z}\right)\bar{a}\theta B_1(x)\varphi_{0,1}(z)$$

$$+\left[\left(\frac{z+\bar{a}r_0(1-z)}{z}\right)(\bar{v}+vz)\theta B_1(x)-(a\bar{b}z+ab+\bar{a})\right]\varphi_{1,1}(z)$$

$$+\left(\frac{z+\bar{a}r_0(1-z)}{z}\right)\theta B_1(x)\varphi_{2,1}(z)+\left(\frac{z-1}{z}\right)r_0\theta aB_1(x)P(0), \qquad (6.14)$$

$$\frac{x-(a\bar{b}z+ab+\bar{a})}{x}\varphi_2(x,z)=(1-z)\bar{a}\bar{\theta}B_2(x)\varphi_{0,1}(z)$$

$$+(z+(1-z)\bar{a}r_0)(\bar{v}+vz)\bar{\theta}B_2(x)\varphi_{1,1}(z)$$

$$+[(z+(1-z)\bar{a}r_0)\bar{\theta}B_2(x)-(a\bar{b}z+ab+\bar{a})]\varphi_{2,1}(z)+(z-1)r_0a\bar{\theta}B_2(x)P(0). \quad (6.15)$$

Setting $x=\bar{a}$ in Eq. (6.11) and $x=a\bar{b}z+ab+\bar{a}$ in Eqs. (6.14) and (6.15), we get

$$a[L(\bar{a})-r_0]P(0)=-\bar{a}\varphi_{0,1}(z)+\bar{a}[\bar{v}+vz][L(\bar{a})-r_0]\varphi_{1,1}(z)$$

$$+\bar{a}[L(\bar{a})-r_0]\varphi_{2,1}(z), \qquad (6.16)$$

$$(1-z)r_0\theta aB_1(a\bar{b}z+ab+\bar{a})P(0)=(1-z)\bar{a}\theta B_1(a\bar{b}z+ab+\bar{a})\varphi_{0,1}(z)$$

$$+(z+(1-z)\bar{a}r_0)\theta B_1(a\bar{b}z+ab+\bar{a})\varphi_{2,1}(z)$$

$$+\left[(z+(1-z)\bar{a}r_0)(\bar{v}+vz)\theta B_1(a\bar{b}z+ab+\bar{a})-z(a\bar{b}z+ab+\bar{a})\right]\varphi_{1,1}(z), \quad (6.17)$$

$$a(1-z)r_0\bar{\theta}B_2(a\bar{b}z+ab+\bar{a})P(0)=(1-z)\bar{a}\bar{\theta}B_2(a\bar{b}z+ab+\bar{a})\varphi_{0,1}(z)$$

$$+(z+(1-z)\bar{a}r_0)(\bar{v}+vz)\bar{\theta}B_2(a\bar{b}z+ab+\bar{a})\varphi_{1,1}(z)$$

$$+[(z+(1-z)\bar{a}r_0)\bar{\theta}B_2(a\bar{b}z+ab+\bar{a})-(a\bar{b}z+ab+\bar{a})]\varphi_{2,1}(z). \qquad (6.18)$$

Using mathematical manipulations the generating functions are obtained as

$$\varphi_{0,1}(z)=\frac{az[L(\bar{a})-r_0][(a\bar{b}z+ab+\bar{a})-(\bar{v}+vz)\theta B_1(a\bar{b}z+ab+\bar{a})-\bar{\theta}zB_2(a\bar{b}z+ab+\bar{a})]}{\Omega(z)}$$

$$\frac{P(0)}{\bar{a}},$$

$$\qquad (6.19)$$

$$\varphi_{1,1}(z) = \frac{aL(\bar{a})(1-z)\theta B_1(\overline{abz} + ab + \bar{a})}{\Omega(z)} P(0), \qquad (6.20)$$

$$\varphi_{2,1}(z) = \frac{aL(\bar{a})(1-z)\bar{\theta}zB_2(\overline{abz} + ab + \bar{a})}{\Omega(z)} P(0) \qquad (6.21)$$

where $\Omega(z) \equiv [(\bar{v} + vz)\theta B_1(\overline{abz} + ab + \bar{a}) + \bar{\theta}zB_2(\overline{abz} + ab + \bar{a})][z + (1-z)\bar{a}L(\bar{a})] - z(\overline{abz} + ab + \bar{a})$.

Lemma 1

1. The   inequality $[\theta B_1(\overline{abz} + ab + \bar{a})(\bar{v} + vz) + \bar{\theta}zB_2(\overline{abz} + ab + \bar{a})]$
$[z + (1-z)\bar{a}L(\bar{a})] - z(\overline{abz} + ab + \bar{a}) > 0$ holds for $0 \le z \le 1$ if and only if
$(\theta \rho_1 + \bar{\theta}\rho_2)\bar{q} + \theta v + \bar{\theta} < \overline{ab} + \bar{a}L(\bar{a})$.
2. The following limit exists if and only if $(\theta \rho_1 + \bar{\theta}\rho_2)\bar{b} + \theta v + \bar{\theta} < \overline{ab} + \bar{a}L(\bar{a})$,

$$\lim_{z \to 1} \frac{(\overline{abz} + ab + \bar{a}) - \theta B_1(\overline{abz} + ab + \bar{a}) - \bar{\theta}zB_2(\overline{abz} + ab + \bar{a})}{\Omega(z)}$$

$$= \frac{\bar{\theta} + \theta v + \bar{b}(\theta \rho_1 + \bar{\theta}\rho_2 - a)}{\overline{ab} + \bar{p}L(\bar{a}) - \bar{b}(\theta \rho_1 + \bar{\theta}\rho_2) - \theta v - \bar{\theta}} \qquad (6.22)$$

$$\lim_{z \to 1} \frac{(1-z)}{\Omega(z)} = \frac{1}{\overline{ab} + \bar{a}L(\bar{a}) - \bar{b}(\theta \rho_1 + \bar{\theta}\rho_2) - \theta v - \bar{\theta}}. \qquad (6.23)$$

Proof. Define:

$$u(z) = [\theta(\bar{v} + vz)B_1(\overline{abz} + ab + \bar{a}) + \bar{\theta}zB_2(\overline{abz} + ab + \bar{a})][z + (1-z)\bar{a}L(\bar{a})],$$

$$v(z) = z(\overline{abz} + ab + \bar{a}).$$

We find the derivatives of the function $u(z)$ and $v(z)$ at $z = 1$ to examine the slope of their tangents.

$$u'(1) = \theta B_1'(1)\overline{ab} + \bar{\theta}B_2'(1)\overline{ab} + \bar{\theta} + 1 - \bar{a}L(\bar{a}) = \bar{b}(\theta \rho_1 + \bar{\theta}\rho_2) + \bar{\theta} + 1 - \bar{a}L(\bar{a}),$$

$$v'(1) = 1 + \overline{ab}.$$

Here, it can be observed that $u'(1) < v'(1)$. Therefore, we have $u(z) > v(z)$ in $0 \le z \le 1$ since u and v are convex functions. Both the limits in statement (2) exist provided denominator is nonzero – i.e., $(\theta \rho_1 + \bar{\theta}\rho_2)\bar{b} + \bar{\theta} < \overline{ab} + \bar{a}L(\bar{a})$. Finally, the theorem given below discusses the solutions of the stationary equations.

Theorem 1

The PGFs for the steady-state distribution of the Markov chain are given by

$$\varphi_0(x,z) = \frac{[L(x) - L(\bar{a})]}{x - \bar{a}}$$

$$\frac{axz[(a\bar{b}z + ab + \bar{a}) - \theta(\bar{v} + vz)B_1(a\bar{b}z + ab + \bar{a}) - \bar{\theta}zB_2(a\bar{b}z + ab + \bar{a})]}{\Omega(z)} P(0),$$

$$(6.24)$$

$$\varphi_1(x,z) = \frac{B_1(x) - B_1(a\bar{b}z + ab + \bar{a})}{x - (a\bar{b}z + ab + \bar{a})} \frac{ax(1-z)\theta L(\bar{a})(a\bar{b}z + ab + \bar{a})}{\Omega(z)} P(0), \qquad (6.25)$$

$$\varphi_2(x,z) = \frac{B_2(x) - B_2(a\bar{b}z + ab + \bar{a})}{x - (a\bar{b}z + ab + \bar{a})} \frac{axz(1-z)\bar{\theta}L(\bar{a})(a\bar{b}z + ab + \bar{a})}{\Omega(z)} P(0), \qquad (6.26)$$

where

$$P(0) = \frac{a\bar{b} + \bar{a}L(\bar{a}) - \bar{b}[\theta\rho_1 + \bar{\theta}\rho_2] - \theta v - \bar{\theta}}{L(\bar{a})[\theta\bar{v} + b[\theta\rho_1 + \bar{\theta}\rho_2 - a]]}. \qquad (6.27)$$

Proof. Substituting Eqs. (6.19)–(6.21) into Eqs. (6.11), (6.14) and (6.15) we derived the generating functions $\varphi_j(x,z), j = 0,1,2$ as given in the statement of the theorem.

Making use of $P(0) + \varphi_0(1,1) + \varphi_1(1,1) + \varphi_2(1,1) = 1$, we calculate $P(0)$ as:

$$P(0) = \frac{a\bar{b} + \bar{a}L(\bar{a}) - \bar{b}[\theta\rho_1 + \bar{\theta}\rho_2] - \theta v - \bar{\theta}}{L(\bar{a})[\theta\bar{v} + b[\theta\rho_1 + \bar{\theta}\rho_2 - a]]}.$$

Making use of the above theorem, we obtain the results stated in the Corollary 1.

Corollary 1

1. The marginal PGF for the orbit size given that the server is free:

$$P(0) + \varphi_0(1,z) =$$

$$\frac{L(\bar{a})[(\bar{a} + az)(\theta B_1(a\bar{b}z + ab + \bar{a})(\bar{v} + vz) + \bar{\theta}zB_2(a\bar{b}z + ab + \bar{a})) - z(a\bar{b}z + ab + \bar{a})]}{\Omega(z)} P(0).$$

$$(6.28)$$

2. The marginal PGF for the orbit size given that the server is serving a customer:

$$\varphi_1(1,z) = \frac{A(\bar{a})\theta(a\bar{b}z + ab + \bar{a})(1 - B_1(a\bar{b}z + ab + \bar{a}))}{\bar{b}\Omega(z)} P(0). \qquad (6.29)$$

3. The marginal PGF for the orbit size given that the server is under repair:

$$\varphi_2(1,z) = \frac{A(\bar{p})\bar{\theta}(p\bar{q}z + pq + \bar{p})z(1 - B_2(p\bar{q}z + pq + \bar{p}))}{\bar{q}\Omega(z)} P(0). \qquad (6.30)$$

4. The PGF of the orbit size:

$$\Psi(z) = P(0) + \varphi_0(1,z) + \varphi_1(1,z) + \varphi_2(1,z) \qquad (6.31)$$

$$= \frac{L(\bar{a})\{\theta(a\bar{b}z + ab + \bar{a})(1-z)(1-vB_1(a\bar{b}z + ab + \bar{a})) + b[a\theta(1-z) +}{z(a\bar{b}z + ab + \bar{a}) - \theta B_1(a\bar{b}z + ab + \bar{a})(a + \bar{a}(\bar{v} + vz)) - \bar{\theta}zB_2(a\bar{b}z + ab + \bar{a})]\}}{\bar{b}\Omega(z)} P(0).$$

5. The PGF for the system size:

$$\varphi(z) = P(0) + \varphi_0(1,z) + z\varphi_1(1,z) + \varphi_2(1,z)$$

$$= \frac{L(\bar{a})\{(a\bar{b}z + \bar{a})\theta(1-z)\bar{v}B_1(a\bar{b}z + ab + \bar{a}) - b[\theta(\bar{a}\bar{v} + \bar{a}vz + az)}{B_1(a\bar{b}z + ab + \bar{a}) + \bar{\theta}zB_2(a\bar{b}z + ab + \bar{a}) - z(a\bar{b}z + ab + \bar{a})]\}}{\bar{b}\Omega(z)} P(0). \qquad (6.32)$$

Using Eqs. (6.24)–(6.27), we derive Eqs. (6.28)–(6.30) and from those results after simple mathematics we obtain Eq. (6.32).

Remark: The M/G/1 retrial queue with feedback and starting failures discussed by Krishnakumar *et al.* (2002) can be approximated by our discrete time retrial queue under study. Suppose that the time is divided into intervals of equal length Δ. The parameters for continuous time system can be approximated as:

$$a = \lambda\Delta, \qquad \beta_{1,1} = 1 - \beta_1\Delta, \qquad \beta_{2,1} = \phi_1\Delta.$$

Taking $b = 0$ (no balking), our results agree with the continuous time counterpart (see, Krishnakumar et al., 2002).

6.3.1 Performance Measures

Here, we present a few important quantities of interest that we derived for the system under consideration.

1. Probability that the system is free:

$$P(0) = \frac{a\bar{b} + \bar{a}L(\bar{a}) - \bar{b}[\theta\rho_1 + \bar{\theta}\rho_2] - \theta v - \bar{\theta}}{L(\bar{a})[\theta\bar{v} + b[\theta\rho_1 + \bar{\theta}\rho_2 - a]]}.$$

2. Probability that the system is occupied:

$$\varphi_0(1,1) + \varphi_1(1,1) + \varphi_2(1,1)$$

$$= \frac{[\theta\rho_1 + \bar{\theta}\rho_2 - a]\bar{b} + \theta v + \bar{\theta} + L(\bar{a})[\bar{v}\theta - (1 - a\bar{b})] + bL(\bar{a})(\theta\rho_1 + \bar{\theta}\rho_2)}{L(\bar{a})[\theta\bar{v} + b[\theta\rho_1 + \bar{\theta}\rho_2 - a]]}.$$

3. Probability of the server being free:

$$P(0) + \varphi_0(1,1) = \frac{\theta\bar{v} - ab - \bar{b}[\theta\rho_1 + \bar{\theta}\rho_2]}{\theta\bar{v} + b[\theta\rho_1 + \bar{\theta}\rho_2 - a]}.$$

4. Probability of the server being busy:

$$\varphi_1(1,1) = \frac{\theta\rho_1}{\theta\bar{v} + b[\theta\rho_1 + \bar{\theta}\rho_2 - a]}.$$

5. Probability of the server is under repair:

$$\varphi_2(1,1) = \frac{\bar{\theta}\rho_2}{\theta\bar{v} + q[\theta\rho_1 + \bar{\theta}\rho_2 - p]}.$$

6. Expected orbit size:

$$E(N) = \Psi'(1)$$

$$= \frac{\{a\bar{b} + \bar{a}L(\bar{a}) - \bar{b}[\theta\rho_1 + \bar{\theta}\rho_2] - \theta v - \bar{\theta}\}F(z) + \bar{b}\{\bar{v}\theta + b(\theta\rho_1 + \bar{\theta}\rho_2 - a)\}G(z)}{2[a\bar{b} + \bar{a}L(\bar{a}) - \bar{b}[\theta\rho_1 + \bar{\theta}\rho_2] - \theta v - \bar{\theta}][\theta\bar{v} + b[\theta\rho_1 + \bar{\theta}\rho_2 - a]]}$$

where

$$F(z) = 2a\bar{b}\theta\bar{v} - 2\theta v\rho_1\bar{b}(\bar{a} + a\bar{b}) + b\{2\bar{b}[\theta\rho_1\bar{a}v + \bar{\theta}\rho_2 - a] + (\bar{b}a)^2[\theta\beta_{1,2} + (\bar{\theta})\beta_{2,2}]\},\}$$

$$G(z) = 2[1 - \bar{a}L(\bar{a})][\bar{b}[\theta\rho_1 + \bar{\theta}\rho_2] + \theta v + \bar{\theta}] + (\bar{b}a)^2[\theta\beta_{1,2} + (\bar{\theta})\beta_{2,2}]$$

$$+ 2\bar{b}[\theta\rho_1 v + \bar{\theta}\rho_2 - a].$$

7. Expected system size:

$$E(L) = \varphi'(1)$$

$$= \frac{\{a\bar{b} + \bar{a}L(\bar{a}) - \bar{b}[\theta\rho_1 + \bar{\theta}\rho_2] - \theta v - \bar{\theta}\}H(z) + \{\theta v(\bar{a} + a\bar{b}) + b\{\bar{b}(\theta\rho_1 + \bar{\theta}\rho_2) - \theta\bar{a}(1 + v) - a\bar{b}\}\}I(z)}{2[a\bar{b} + \bar{a}L(\bar{a}) - \bar{b}[\theta\rho_1 + \bar{\theta}\rho_2] - \theta v - \bar{\theta}][\theta\bar{v} + b[\theta\rho_1 + \bar{\theta}\rho_2 - a]]},$$

where

$$H(z) = 2ab\bar{v}\theta + 2\bar{b}\theta\bar{v}\rho_1(\bar{a} + ab) + b\{2(a + \bar{a}v)\theta\rho_1\bar{b}$$
$$+ (\bar{b}a)^2[\theta\beta_{1,2} + (\bar{\theta})\beta_{2,2}] + 2\bar{\theta}\bar{b}\rho_2 - 2a\bar{b}\},$$

$$I(z) = 2[1 - \bar{a}L(\bar{a})][\bar{b}[\theta\rho_1 + \bar{\theta}\rho_2] + \theta v + \bar{\theta}] + (\bar{b}a)^2[\theta\beta_{1,2}$$
$$+ (\bar{\theta})\beta_{2,2}] + 2\bar{\theta}\bar{b}\rho_2 - 2a\bar{b} + 2\theta v\rho_1\bar{b}.$$

8. Expected waiting time:

$$W = \frac{E(L)}{a}.$$

6.3.2   DEDUCTIONS

We deduced a few earlier results as the special case of model proposed in this chapter.
   Case (i): As we let $\bar{b} = 1, \varphi(z)$ reduces to the expression

$$\varphi(z) = \frac{L(\bar{a})\{(az + \bar{a})\theta(1 - z)B_1(az + \bar{a})}{[(\bar{v} + vz)\theta B_1(az + \bar{a}) + \bar{\theta}zB_2(az + \bar{a})][z + (1 - z)\bar{a}L(\bar{a})] - z(az + \bar{a})}P(0),$$

where

$$P(0) = \frac{a + \bar{a}L(\bar{a}) - [\theta\rho_1 + \bar{\theta}\rho_2 + v\theta + \bar{\theta}]}{L(\bar{a})\theta\bar{v}}.$$

The above is the PGF of the system size for the system with starting failure and feedback of customers, which coincides with [17].
   Case (ii): When $\bar{v} = 1, \varphi(z)$ reduces to

$$\varphi(z) =$$

$$\frac{L(\bar{a})\{(a\bar{b}z + \bar{a})\theta(1 - z)B_1(a\bar{b}z + ab + \bar{a}) - b[\theta(\bar{a} + az)B_1(a\bar{b}z + ab + \bar{a})}{\bar{b}\{[\theta B_1(a\bar{b}z + ab + \bar{a}) + \bar{\theta}zB_2(a\bar{b}z + ab + \bar{a})][z + (1 - z)\bar{a}L(\bar{a})] - z(a\bar{b}z + ab + \bar{a})\}}P(0),$$
$$\text{with numerator continuation } +\bar{\theta}zB_2(a\bar{b}z + ab + \bar{a}) - z(a\bar{b}z + ab + \bar{a})]\}$$

where

$$P(0) = \frac{a\bar{b} + \bar{a}L(\bar{a}) - [\theta\rho_1 + \bar{\theta}\rho_2 + \bar{\theta}]}{L(\bar{a})\theta + b[\theta\rho_1 + \bar{\theta}\rho_2 - a]}$$

The above is the PGF of the system size in "the standard $Geo/G/1/\infty$ queue with starting failures and balking customers".

Case (iii): When $\bar{v} = 1$, $\varphi(z)$ has the expression

$$\varphi(z) = \frac{\begin{aligned}&(a\bar{b}z+\bar{a})\theta(1-z)B_1(a\bar{b}z+ab+\bar{a})-b\{[\theta(\bar{a}+az)B_1(a\bar{b}z+ab+\bar{a})+\\&\bar{\theta}zB_2(a\bar{b}z+ab+\bar{a})-z(a\bar{b}z+ab+\bar{a})]\}\end{aligned}}{b\{[\theta B_1(a\bar{b}z+ab+\bar{a})+\bar{\theta}zB_2(a\bar{b}z+ab+\bar{a})][\bar{a}+az]-z(a\bar{b}z+ab+\bar{a})\}} \cdot \frac{\theta-ab-\bar{b}[\theta\rho_1+\bar{\theta}\rho_2]}{\theta+b[\theta\rho_1+\bar{\theta}\rho_2-a]}.$$

This is the PGF for the system size in "the *Geo / G / 1* queue with impatient customers and server subjected to starting failures".

Case (iv): When $\theta = \bar{b} = \bar{v} = 1$, our model reduces to "*Geo / G / 1* retrial queue with reliable server" and we have

$$\varphi_0(x,z) = \frac{[L(x)-L(\bar{a})]}{x-\bar{a}} \frac{axz[(az+\bar{a})-B_1(az+\bar{a})]}{(1-z)\bar{a}L(\bar{a})B_1(az+\bar{a})-z[(az+\bar{a})-B_1(az+\bar{a})]} P(0),$$

$$\varphi_1(x,z) = \frac{B_1(x)-B_1(az+\bar{a})}{x-(az+\bar{a})} \frac{(1-z)axL(\bar{a})(az+\bar{a})}{(1-z)\bar{a}L(\bar{a})B_1(az+\bar{a})-z[(az+\bar{a})-B_1(az+\bar{a})]} P(0),$$

with

$$P(0) = \frac{a+\bar{a}L(\bar{a})-\rho_1}{L(\bar{a})},$$

which are exactly the generating functions of Theorem 1 in Atencia and Moreno (2004a).

6.4　STOCHASTIC DECOMPOSITION PROPERTY

For the system under consideration, we establish the decomposition of the random variable representing the system size in two different ways.

Theorem 2

1. The number of customers L, in the system can be written as $L = L' + M'$, where L' is the number of customers in "the standard *Geo / G / 1* queue with impatient customers and starting failures" and M' is the number of retrial customers when the server is free.
2. The system size L can also be rewritten as $L = L'' + M''$, where L'' is the system size in "the *Geo / G / 1* queue with impatient customers" and M'' is the number of customers retrying for service when the server is free or under repair.

Proof. We can easily rewrite equation (6.32) in the following two different ways.

$$\varphi(z) = \frac{\begin{array}{l}(a\bar{b}z + \bar{a})\theta(1-z)\bar{v}B_1(a\bar{b}z + ab + \bar{a}) - b\{\theta(\bar{a}v + \bar{a}vz + az)B_1(a\bar{b}z + ab + \bar{a})\\ +\bar{\theta}zB_2(a\bar{b}z + ab + \bar{a}) - z(a\bar{b}z + ab + \bar{a})\}\{\theta - ab - \bar{b}[\theta\rho_1 + \bar{\theta}\rho_2]\}\end{array}}{\begin{array}{l}\bar{b}\{[\theta B_1(a\bar{b}z + ab + \bar{a}) + \bar{\theta}zB_2(a\bar{b}z + ab + \bar{a})][\bar{a} + az]\\ -z(a\bar{b}z + ab + \bar{a})\}\{\theta\bar{v} + b[\theta\rho_1 + \bar{\theta}\rho_2 - a]\}\end{array}}$$

$$\times \frac{P(0) + \varphi_0(1,z)}{P(0) + \varphi_0(1,1)}. \tag{6.33}$$

Here, the first term is the PGF of the system size in "*Geo / G /* 1 queue with starting failure" and the second term of the product is the PGF of the number of retrial customers when the server is free.

$$\varphi(z) = \frac{\begin{array}{l}(a\bar{b}z + \bar{a})(1-z)\bar{v}B_1(a\bar{b}z + ab + \bar{a}) - b\{(\bar{a}v + \bar{a}vz + az)B_1(a\bar{b}z + ab + \bar{a})\\ -z(a\bar{b}z + ab + \bar{a})\}(1 - ab - \bar{b}\rho_1)\end{array}}{\bar{b}\{B_1(a\bar{b}z + ab + \bar{a})[\bar{a} + az] - z(a\bar{b}z + ab + \bar{a})\}(\bar{v} + b[\rho_1 - a])}$$

$$\times \frac{P(0) + \varphi_0(1,z) + \varphi_2(1,z)}{P(0) + \varphi_0(1,1) + \varphi_2(1,1)}. \tag{6.34}$$

In this representation of $\varphi(z)$, the first term of the product is the PGF of the system size in *Geo / G /* 1 standard queue and the other term is the PGF of the number of customers retrying their service when the server is idle or under repair.

6.5   NUMERICAL ILLUSTRATION

Exhaustive numerical results are presented in this section to describe the behavior of the queueing systems under study. We examine the effect of the system parameters such as arrival probability (a), balking probability (b), feedback probability (v) and the probability that the server is started (θ) to serve the customer successfully on some of the crucial performance measures of our model. The values of the parameters are chosen appropriately and the numerical results are obtained by using MATLAB software.

Figure 6.1 $(a) - (l)$ depicts the impact of a, b, θ and v on $P(0)$. As expected intuitively, $P(0)$, the probability that the system is empty, decreases with the increase in a for any values of b, θ and v (Figure 6.1$(a) - (c)$). Further, $P(0)$ is almost stable for increasing values of b. We can see from Figure 6.1$(d) - (f)$, $P(0)$ remains unaltered as for as b is concerned but it is higher for higher values of a, θ and also v. Figure 6.1 $(g) - (i)$ shows an increasing trend for $P(0)$ for increasing values of θ. Figure 6.1 (h) confirms the observation that $P(0)$ is not much affected by b. $P(0)$ is plotted against v for different values of a, b and θ in Figure 6.1 $(j) - (l)$. We can see from these Figure 6.1 that $P(0)$ decreases with the increment in v. Further, we can see it increases with increase in b and θ but decreases with increase in a.

The effect of a, b, θ and v on $E(L)$ is plotted in Figure 6.2$(a) - (l)$. The behavior of E(L) with respect to a is described in Figure 6.2$(a) - (c)$. It is noticed that the

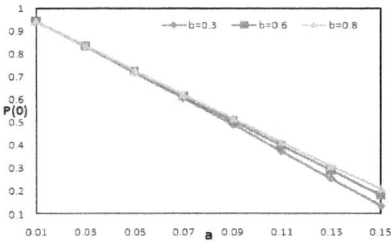

(a) $P(0)$ versus a for $b = 0.3, 0.6, 0.8$

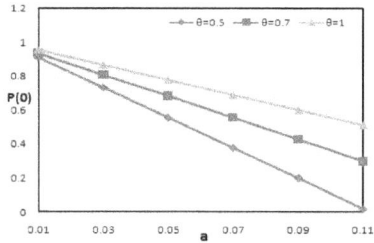

(b) $P(0)$ versus a for $\theta = 0.5, 0.7, 1$

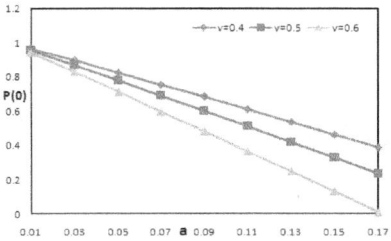

(c) $P(0)$ versus a for $v = 0.4, 0.5, 0.6$

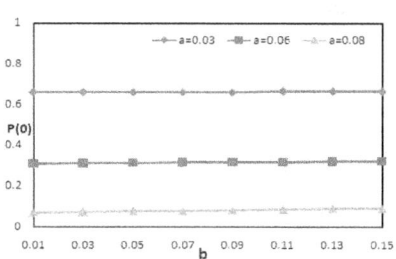

(d) $P(0)$ versus b for $a = 0.03, 0.06, 0.08$

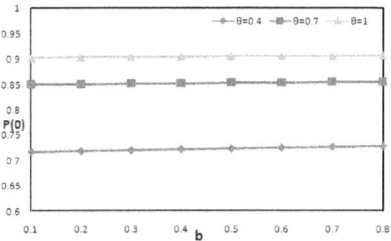

(e) $P(0)$ versus b for $\theta = 0.4, 0.7, 1$

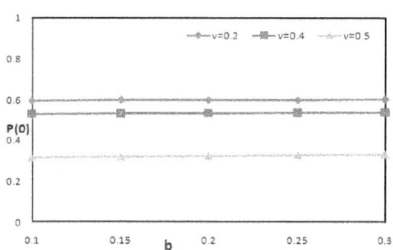

(f) $P(0)$ versus b for $v = 0.2, 0.4, 0.5$

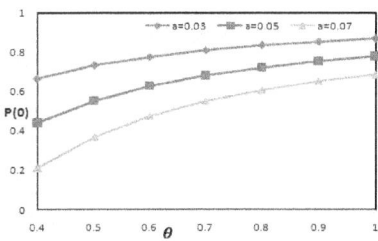

(g) $P(0)$ versus θ for $a = 0.03, 0.05, 0.07$

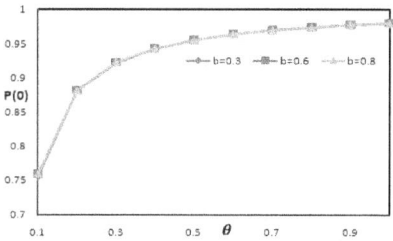

(h) $P(0)$ versus θ for $b = 0.3, 0.6, 0.8$

FIGURE 6.1 Impact of a, b, θ and v on $P(0)$.

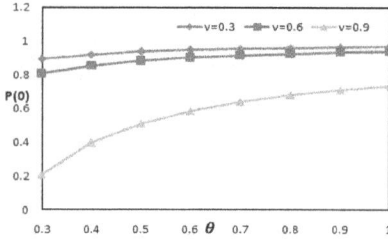

(i) $P(0)$ versus θ for $v = 0.3, 0.6, 0.9$

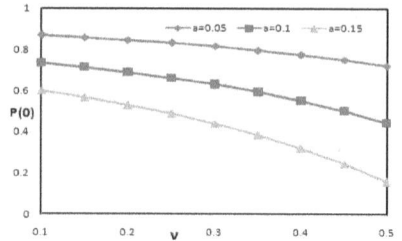

(j) $P(0)$ versus v for $a = 0.05, 0.1, 0.15$

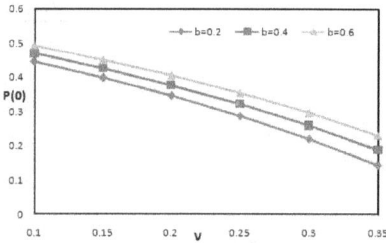

(k) $P(0)$ versus v for $b = 0.2, 0.4, 0.6$

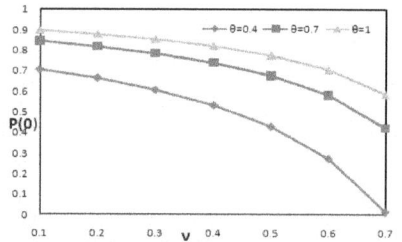

(l) $P(0)$ versus v for $\theta = 0.4, 0.7, 1$

FIGURE 6.1 *(Continued)*

expected system size $E(L)$ increases for increasing values of a which matches with our intuition. Figure $6.2(d) - (f)$ showcases the trend of $E(L)$ for increasing values of b. $E(L)$ decreases as b increases. Also it is observed that $E(L)$ is higher for higher values of a and v but it is the other way for θ. $E(L)$ is drawn for increasing values of θ for different values of a, b and v in Figure $6.2(g) - (i)$. As is expected, $E(L)$ decreases for increasing values of θ in all the three different cases. Also, $E(L)$ is higher for the higher values of a and v but is very low for higher values of b. Finally, in Figure $6.2(j) - (l)$, $E(L)$ is depicted for increasing values of v for various values of a, b and θ. We observe in Figure 6.2 an increasing trend for $E(L)$ for increasing v for a, b and θ. Further, $E(L)$ is higher for the higher values of a but it is on the other way for b and θ. Also, we see a different pattern when a is very small ($a = 0.05$) and b is more ($b = 0.6$) which may be the combined effect of v and a (Figure $6.2(j)$) and v and b (Figure 6.2(k)). The numerical results established above demonstrates that our analytical results are valid. Further, they can provide some information required for the concerned system designers for making optimal designs.

6.6   CONCLUSIONS

In the present study, we considered a *Geo / G / 1* queue with retrial customers. We assume that the server is prone to starting failure and the customer may rejoin the system for another service if his service is not satisfactory. Further, we have taken into consideration, the customer's dilemma whether to join the system or leave if he or she anticipates a longer waiting time. We have analyzed the system in the early arrival set up where the departure is having precedence over the arrivals. It

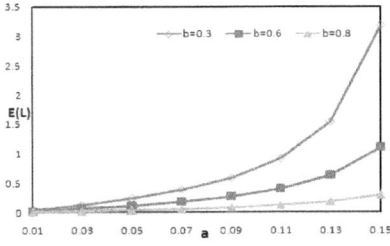

(a) $E(L)$ versus a for $b = 0.3, 0.6, 0.8$

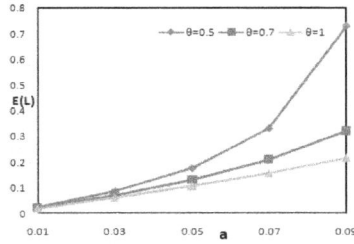

(b) $E(L)$ versus a for $\theta = 0.5, 0.7, 1$

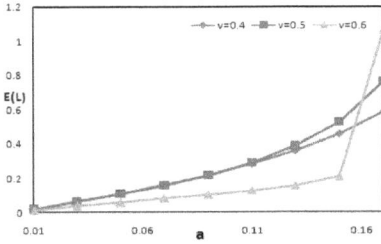

(c) $E(L)$ versus a for $v = 0.4, 0.5, 0.6$

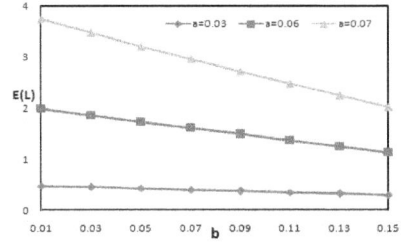

(d) $E(L)$ versus b for $a = 0.03, 0.06, 0.08$

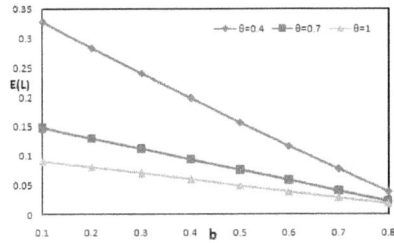

(e) $E(L)$ versus b for $\theta = 0.4, 0.7, 1$

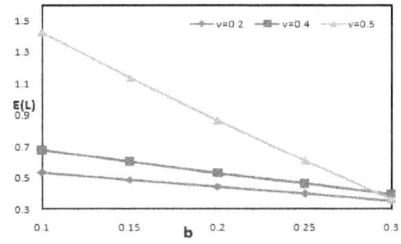

(f) $E(L)$ versus b for $v = 0.2, 0.4, 0.5$

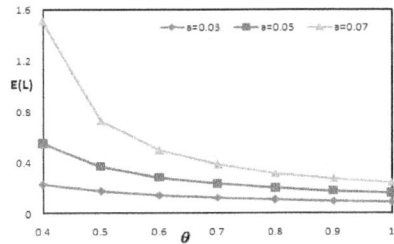

(g) $E(L)$ versus θ for $a = 0.03, 0.05, 0.07$

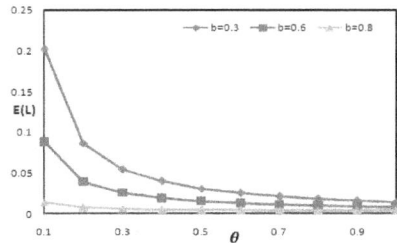

(h) $E(L)$ versus θ for $b = 0.3, 0.6, 0.8$

FIGURE 6.2   Impact of a, b, θ and v on $E(L)$.

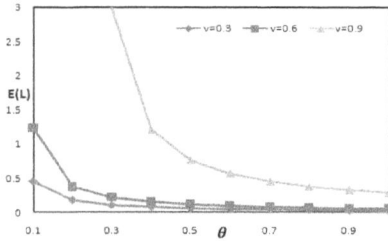

(i) $E(L)$ versus θ for $v = 0.3, 0.6, 0.9$

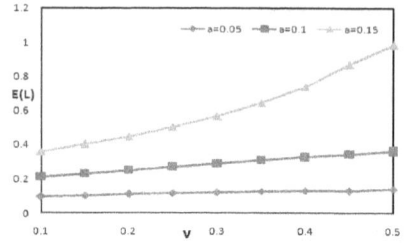

(j) $E(L)$ versus v for $a = 0.05, 0.1, 0.15$

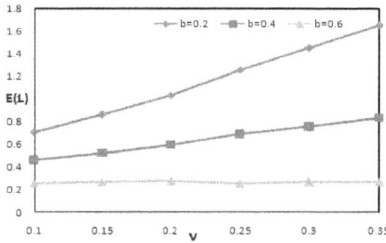

(k) $E(L)$ versus v for $b = 0.2, 0.4, 0.6$

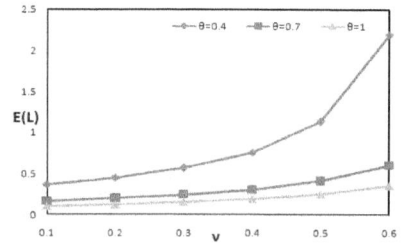

(l) $E(L)$ versus v for $\theta = 0.4, 0.7, 1$

FIGURE 6.2 *(Continued)*

is observed that the $P(0)$ increases for increasing values of θ, the probability that the server is started successfully and decreases with increasing value of feedback probability v. Further, it is noticed that the expected system size increases with the increase in the value of v and decreases with the increase in the values of θ. The trends shown in the graphs are in line with our intuition, which ensures the correctness of the analytical results obtained. Our model can be used to study a flexible manufacturing system where the arrival of components and their service time are occurring at regularly spaced time epochs. The arrival of the components, the manufacturing unit, a pool of waiting components, the failure of the system, unsatisfied components processed again may correspond to the customers, the server, retrial customers, starting failure and feedback, respectively. The future work may include study of the same system in late arrival setup, waiting time distribution and cost analysis considering holding cost and the cost of losing a customer.

ACKNOWLEDGEMENT

The authors thank the referees for their valuable input in improving the manuscript.

REFERENCES

Aboul-Hassan A., Rabia S. I., and Kadry A. 2005. Analytical study of a discrete time retrial queue with balking customers and early arrival scheme. *Alexandria Engineering Journal* 44:919–925.

Aboul-Hassan A., Rabia S. I., and Kadry A. 2008. A discrete time *Geom/G/1* retrial queue with general retrial times and balking customers. *Journal of the Korean Statistical Society* 37:335–348.

Artalejo J. R. 1999. A classified bibliography of research on retrial queues: progress in 1990–1999. *Top* 7:187–211.

Artalejo, J. R. 2010. Accessible bibliography on retrial queues: progress in 2000–2009. *Mathematical and Computer Modelling* 51:1071–1081.

Artalejo J. R., and Gomez-Corral A. 2008. Retrial Queueing Systems: A Computational Approach, Springer, Berlin.

Atencia I., Ruiz I. F. and Sanchez S. 2009. A discrete-time retrial queueing system with starting failures, Bernoulli feedback and general retrial times. *Computers & Industrial Engineering* 57(4): 1291–1299.

Atencia, I. 2017. A *Geo/G/1* retrial queueing system with priority services, *European Journal of Operational Research.* 256:178–186.

Atencia I., and Moreno P. 2004a. A discrete-time *Geo/G/1* retrial queue with general retrial times. *Queueing Systems* 48:5–21.

Atencia I., and Moreno P. 2004b. A discrete-time *Geo/G/1* retrial queue with optional service. *Proceedings of the Fifth International Workshop on Retrial Queues* 9:117–121.

Atencia I., and Moreno P. 2006. A discrete-time *Geo/G/1* retrial queue with the server subject to starting failures. *Annals of Operational Research.* 141:85–107.

Bruneel, H. and Kim B.G. 1993. Discrete-Time Models for Communication Systems Including ATM. Kluwer Academic Publishers, Boston.

Choi B. D., and Kim J. W. 1997. Discrete-time Geo1,Geo2/G/1 retrial queueing system with two types of calls. *Computers and Mathematics with Applications* 33:79–88.

Gao S., and Wang X. 2019. Analysis of a single server retrial queue with server vacation and two waiting buffers based on ATM networks. *Mathematical Problems in Engineering* 1–14. https://doi.org/10.1155/2019/4193404.

Gomez-Corral A. 2006. A bibliographical guide to the analysis of the retrial queues through matrix analytic techniques. *Annals of Operations Research* 141:163–191.

Krishna Kumar B., Pavai Madheswari S. and Vijayakumar A. 2002. The *M/G/1* rertial queue with feedback and starting failures. *Applied Mathematical Modelling* 26:1057–1075.

Kulshrestha, R. and Shruti 2020. Discrete-Time Analysis of Communication Networks with Second Optional Service and Negative User Arrivals. In Lecture Notes in Networks and Systems. Springer during ICCIS 2019 at SKIT. Jaipur, India.

Lan S., and Tang Y. 2019. Performance analysis of a discrete-time *Geom/G/1* retrial queue with non-preemptive priority, working vacations and vacation interruption. *Journal of Industrial & Management Optimization* 15:1421–1446.

Madan, K.C. and AI-Rawwash, M. 2005. On the M X /G/1 queue with feedback and optional server vacations based on a single vacation policy. *Applied Mathematics and Computation.* 160(3): 909–919.

Meisling T. 1958. Discrete time queueing theory. *Operations Research* 6:96–105.

Nobel R. 2016. A mixed discrete time delay/retrial queueing model for handover calls competing for a target channel. *Queueing Theory and Network Applications* 383:173–185.

Pavai Madheswari S., Krishna Kumar B., and Suganthi P. 2019. Analysis of *M/G/1* retrial queues with second optional service and customer balking under two types of Bernoulli vacation schedule. *RAIRO-Operational Research* 53:415–443.

Takacs L. 1963. The Stochastic law of the busy period for a single server queue with poisson input. *Journal of Mathematical Analysis and applications* 6:33–42.

Takagi H. 1993. Queueing Analysis: A Foundation of Performance Evaluation, Discrete-Time Systems. North-Holland, Amsterdam.

Upadhyaya S. 2018. Performance analysis of a discrete-time *Geom/G/1* retrial queue under J-vacation policy. *International Journal of Industrial and Systems Engineering* 29:369–388.

Wang J., and Zhao Q. 2007. A discrete time *Geom/G/1* retrial queue with starting failures and second optional service. *Computers & Mathematics with Applications* 53:115–127.

Wei C., Cai L. and Wang J. A. 2015. A discrete time *Geom/G/1* retrial queue with balking customers and second optional service. *OPSEARCH* 53:344–357.

Woodward M.E. 1994. Communication and Computer Networks: Modelling with Discrete-Time Queues, IEEE Computer Society Press, Los Alamitos, CA.

Wu J.B., Wang J.X., and Liu Z.M. 2013. A discrete time *Geom/G/1* retrial queue with preferred and impatient customers. *Applied Mathematical Modeling*. 37:2552–2561.

7 Application of Optimization and Statistical Techniques in Post-Harvest Supply Chain

A Systematic Literature Survey

Rahul Priyadarshi
Birla Institute of Technology & Science, Pilani, Rajasthan, India
GITAM School of Business – Bengaluru, India

Srikanta Routroy, Girish Kant Garg
Birla Institute of Technology & Science, Pilani, Rajasthan, India

CONTENTS

7.1 INTRODUCTION

Fresh produce is an important part of life. The post-harvest supply chain (PHSC) focuses on the journey of produce from yielding in a farm to cleaning, grading, packaging, handling, transportation, storage, distribution, food-processing, retail and exports, etc. The PHSCs are practiced across the globe to deliver various fresh produces to the end customer. The PHSCs are often referred to as fresh supply chains, food supply chains and cold chains etc. The different PHSCs for various produce require a specific level of packaging, handling, ventilated/refrigerated storage during transportation to retain quality and yield best values. It is also well known that one-third of the annually grown produce worldwide are spoilt at various PHSC junctions. The PHSC losses occur mainly due to microbial damage (Kereth et al., 2013), physical damage (Barker et al., 2017) and temperature abuse (Aung and Chang, 2014; Ndraha et al., 2018), etc. Efforts are continuously made by researchers to make the PHSCs responsive, cost-efficient, quality-efficient and sustainable. The optimization of PHSCs have become a crucial task with a sequence of depleting natural resources and an increasing carbon footprint. Thus, the significant issues in the PHSCs are to be addressed with the help of statistical, empirical and operational approaches. The PHSC losses occurring at every junction of the chain have been assessed for various produce in the literature to mitigate the financial and produce losses. Although, finding the voids between existing studies, modernization, real-life practice and implementations is a persistent process. There are a total of 8,038 cold storages in India, 1,540 godowns associated with Food Corporation of India, and 1,831 warehouses available with state warehousing corporations in India across the nation (includes privately owned, cooperatives and government regulated) (Indiastat, 2019a). There's 76.15 lakh tonnes of warehousing capacity available in India provided by Central Warehousing Corporation in India (Indiastat, 2019b). India is the nation with second longest road network in the world i.e. 4,320,000 km (Walkthroughindia.com, 2019). The Indian railway comprises the fourth longest route length in the world i.e. 67,368 km (Wikipedia, 2019). The annual demand for domestic air freight is predicted to become 1.1 million tonnes by the year 2025 in India (*Economic Times*, 2019). Thus, there is a lot of potential to efficiently plan and research over warehousing, routing and mode of transportation for time-bound commodities.

7.2 LITERATURE REVIEW

The PHSC loss situation is observed by the researchers since ages when the term "post-harvest losses" was addressed by Bourne (1977). The theoretical, experimental, analytical, exploratory, empirical research articles are based on optimization models and statistical studies. The post-harvest loss situation, in-house utility and produce export issues are reported across the world, and that includes developed nations, developing nations, underdeveloped nations, cold-climate nations, arid nations, etc. In the recent times, the causal factors leading to PHSC losses are highlighted by an array of researchers (Gardas et al., 2017; Raut et al., 2018; Bendinelli et al., 2019). The following review methodology explains the process adapted to perform the study.

7.2.1 REVIEW METHODOLOGY

The literature review methodology includes the segregation of fundamental information from the literature survey covering the key studies presented with empirical, statistical and optimization approaches performed along different stages of supply chain. The researches performed at different stages considered in the work are organized with respect to farmer level, warehousing, transportation and handling, manufacturing and processing, markets and retail, and supply chain as a whole. The key studies are tabulated in Table 7.1 below. The table highlights stage-wise studies performed by an array of researchers to reduce PHSC losses:

7.3 APPLICATION OF OPTIMIZATION AND STATISTICAL TECHNIQUES IN POST-HARVEST SUPPLY CHAIN

The literature thus reveals various researches performed in the past over different stages to address the PHSC loss situation. The points below are the key areas that are addressed by various researchers and hold crucial prospects to be worked upon:

- *Vehicle routing problem:* The vehicle routing problems are worked out to optimize the transportation routes and to minimize number of halts/junctions to reduce delivery time and cycle time for fast-moving consumer goods (FMCG). Optimizing the transportation routes and reducing the distance reduces the chances of produce spoilage and assists in retaining the produce quality. Vehicle routing is a major problem that needs attention at a major scale for a variety of fresh, dairy and meat produce in a developing nation. A few potential applications could be to model the vehicle routing problem for fresh produce transportation from to local and central markets, warehouses, processing plants and exports in the PHSC with integer programming model for reduced transportation (Lamsal et al., 2016), multi-objective programming for distribution and routing (Wang et al., 2016), hybrid optimization models to retain quality with improved logistics design (De Keizer et al., 2015), PSO-VNS (Govindan et al., 2014), GA (Rabbani et al., 2016) and stochastic hybrid modelling to analyze energy consumption in cold chains (Laguerre et al., 2013), etc.
- *Warehouse layout optimization:* The warehouse layout optimization is a technique that attempts to ensure effective usage of available storage volume in a warehouse with organized stacking and systematic allocation of agricultural commodities. The produce to be stored includes seasonal fruits, vegetables (tomato and onion) and staple crops for a definite time. The need of the hour is to establish efficient multiproduct warehouses to store an array of agriculture produce with controlled-temperature divisions according to remaining shelf-life and potential spoilage ratio of the produce at the rural and district levels. The academic and institutional research on modernization of warehouses and its implementation will reduce fresh fruit losses and crop losses significantly. A developing nation such as India can surely reduce its seasonal and annual produce wastage situation with the availability of 11,409 and increasing cold storage and warehousing facilities across the nation. The methods such as queuing theory for storage performance (East, 2011), game theory for grain value chain behavior (Brewin, 2016), optimization-integrated value stream mapping (Trebuna et al., 2019) and golden rectangle fundamentals/generic rectangular floor plans (Shekhawat, 2015; Tinello et al., 2016; Shekhawat, 2018) have already been

TABLE 7.1
Research Contributions and Approaches Adapted

Literature Survey	Contribution and Approach
Farmer Stage	
Kong et al. (2019)	Short term farming: mixed integer linear programming
Kale and Sonavane (2019)	Internet of Things (IoT) based farming optimization: Improved Genetic Algorithm (IGA) – Extreme Learning Machine (ELM)
Priyadarshi and Routroy (2018)	Multi-criteria decision-making problem to select the level of vertical integration at rural stage using Fuzzy-Analytical Hierarchy Process (FAHP)
Heenkenda and Chandrakumara (2016)	Multivariate tests and correlations to study impact on farming performance
Ge et al. (2016)	Stochastic optimization-simulation: sensitivity analysis for futuristic goals and to validate the existing practice
Sheu (2016)	Structural Equation Modelling (SEM): assessed the measures to be taken for recovery during post-disaster situation
Pauls-worm et al. (2015)	Non-linear optimization approach to minimize costs
Ahumada and Villalobos (2009, 2011)	Stochastic programming for cultivation and growing plans
Rong et al. (2011)	Optimization approach: quality degradation model
Warehousing	
Guan et al. (2019) and Liu et al. (2018)	Multifacility layout problem: Particle Swarm Optimization (PSO)
Ansari and Smith (2017)	Warehouse layout simulation: interface for warehousing based on optimal routing
Hamidi et al. (2017)	Location and networks design: reliability analysis (location and demand optimization)
Guan and Lin (2016)	Single row layout problem: Variable Neighborhood Search (VNS) and Ant Colony Optimization (ACO)
Mirmajlesi and Shafaei (2016)	Location-allocation problem: multistage, multiproduct, capacity-networking problem; First-In-First-Out (FIFO) policy
East (2011)	Storage performance: queuing theory and statistical performance analysis; fitting the Weibull distribution model
Xiao et al. (2007)	Layout optimization: hybrid compaction

Transportation and Handling

Springael et al. (2018)	Efficient transportation routes: optimal routing problem
Mendoza et al. (2017)	Post-harvest handling: statistical survey (two-tailed test) for maize
Lamsal et al. (2016)	Handling and transportation: integer programming model for reduced transportation; proposes for storage and processing
Mishra et al. (2016)	Supply chain modification: nonlinear programming approach to optimize storage temperature by taking quality, cost and safety as decision variables for leafy greens
Wang et al. (2016)	Distribution and routing: multi-objective programming for vehicle routing considering spoilage factor
Rabbani et al. (2016)	Vehicle routing problem: multi-objective programming solved with Genetic Algorithm (GA)
Song and Ko (2016)	Optimal vehicle routes: nonlinear model for vehicle routing problem with customer satisfaction and produce freshness as objectives and sensitivity analysis to validate the model
Hyland et al. (2016)	Effective rail transportation: time model; cost model; capacity model
De Keizer et al. (2015)	Logistics network design: hybrid optimization approach focusing on product quality
Govindan et al. (2014)	Vehicle and location routing problem: PSO and VNS
Laguerre et al. (2013)	Energy consumption in cold chain: dynamic and stochastic hybrid modelling

Manufacturing and Processing

Majiwa et al. (2018)	Rice processing efficiency: Data Envelopment Analysis (DEA)
Banasik et al. (2017)	Close-loop supply chain: linear programming model for profit maximization and exergy loss minimization
Pauls-worm et al. (2013)	Production planning with non-uniform demand: stochastic programming approach; FIFO strategy
Wu et al. (2012)	Stochastic production-inventory model for produce-as-per-order strategy and produce-to-stock strategy with multilevel storage conditions and multiple retailers

Markets and Retail

Priyadarshi et al. (2019a)	Forecasting methodology for fresh produce at retail
Mowrey et al. (2018)	Retail store optimization (rack layout): PSO

(Continued)

TABLE 7.1
(Continued)

Literature Survey	Contribution and Approach
Adenso-Díaz et al. (2017)	Dynamic pricing: mathematical model to analyze degradation of food inventory over the time
Dobson et al. (2017)	Economic Order Quantity (EOQ) model for fresh produce considering shelf life and its effect on demand at retail
Malik et al. (2016)	Sensitivity analysis for inventory model to reduce spoilage
Schlosser (2015)	Dynamic pricing: state space; Markov models; Hamilton–Jacobi–Bellman (HJB) equation
Lee et al. (2014)	Inventory planning: revenue maximization with First-In-First-Out (FIFO) and Last-In-First-Out (LIFO) ordering strategies
Chen and Zhou (2014)	Decision making for stochastic demands: optimal order quantity focusing upon loss aversion with customer returns
Sternbeck and Kuhn (2014)	Fault with delivery patterns from distribution to retail leading to productivity losses: Binary integer model
Sarker et al. (2000)	Optimal order policy for perishables subject to inflation, payment delay and stock-outs; order quantity shall be low when there are expected delays; also the order quantity gets influenced with the rate of inflation
Supply Chain as a Whole	
Ambler et al. (2018)	Post-harvest loss assessment: survey based statistical analysis during harvesting, transportation, processing and storage
Ibrahim et al. (2018)	Loss occurrence survey: statistical analysis at different stages for paddy rice and milled rice
Amentae et al. (2017)	Value chain analysis: statistical approach for wheat; loss estimation and Tobit model for loss factor determination
Ansah et al. (2017)	Post-harvest loss management (PLM): formulation of empirical structural models
Mizgier (2017)	Supply chain risks: using universal/total/over-all sensitivity analysis and aggregation with multiple products at a time
Brewin (2016)	Grain value chain behavior: game theory
Heydari et al. (2016)	Lead time aggregation for a three stage supply chain: stochastic model
Nourbakhsh et al. (2016)	Optimal network design for pre-processing facilities and sensitivity analysis to reduce system cost, infrastructure investment and economic cost due to losses
Zhao et al. (2016)	Minimizing the operational costs by obtaining a match between optimal production output and inventory; incorporated seasonality

An and Ouyang (2016)	Supply chain design for losses at transportation, storage and processing: optimization approach to increase profit and reduce post-harvest loss by analyzing grain processing/ storage facility and price for grain purchase
Busch and Spiller (2016)	Non-discriminated revenue distribution policies: exploratory factor analysis and regression study
Kasso and Bekele (2016)	Performed over various factors for fresh produce: statistical analysis
Gao (2015)	Vertical integration: collaborative forecasting
Aung and Chang (2014)	Cold chain temperature management: Wireless Sensor Network (WSN) based quality monitoring; product metabolism and Euclidean distance based cost monitoring
Seyedhosseini and Ghoreyshi (2014)	Optimization problem for production, distribution and routing: vehicle routing problem (with time windows) for cost optimization; PSO
Nagare et al. (2013)	Retail inventory management: two bin strategy and sensitivity analysis
Begum et al. (2012)	Endorses for farmer's education and training, and establishment of warehouses near utility (market/processing): multistage sampling design; logistic regression; economic analysis for losses
Dabbene et al. (2008)	Fresh produce supply chain: optimization approach to improve supply chain performance; dealing with uncertainty; physical and logistic parameters; refrigeration
Ortmann (2005)	Efficiency and competitiveness: three mathematical techniques to model the constraints to enhance efficiency and competitiveness, improvement of usage, better investments and logistics infrastructure sharing

applied to layout problems. The warehouse-wise and region-wise case studies on efficient warehousing will help the society with increased produce availability and reduced spoilage ratio during warehousing.

- *Market and retail layout optimization:* The market and retail layouts are to be improvised to utilize the available space optimally provided within the facilities, such as incoming produce depots, trading spots for farmers and traders, distribution points, vendor space management, aisle design, crowd management, warehousing, waste recycling/disposal facility and exit checkpoints, etc. for markets and retail layouts as per strategic necessity. The layouts are required to be optimized to manage space, ease of moving temporary lots and effective utilization, etc. The various approaches adapted to perform similar studies are value stream mapping to optimize market layouts, PSO for retail store optimization (Mowrey et al., 2018) and multifacility layout problem (Guan et al., 2019), VNS and ACO for single row layout problem (Guan and Lin, 2016), chaotic fruit fly algorithm for layout optimization (CNKI, 2015), and golden rectangle fundamentals/generic rectangular floor plans (Tinello et al., 2016; Shekhawat, 2018), etc.

- *Temperature management (optimization):* The temperature management is to regulate/stabilize the produce temperature during transportation, storage and retail. There are an ample number of studies available that highlight the role of temperature monitoring during transportation, storage and retail coupled with radio-frequency identification (RFID) devices. The innovative researches propose the utility of IT enabled cold storages and cold containers to be used during storage and transportation. The utility of optimization and statistical approaches into the research will serve the domain to reach optimality. The multi-objective programming coupled with wireless sensor networks (Aung and Chang, 2014), multi-objective nonlinear programming to optimize storage temperature (Mishra et al., 2016) and exergy-loss minimization with the help of close-loop supply chain (Banasik et al., 2017) as promising approaches highlighted in Table 7.1.

- *Supply chain cost minimization:* There are a number of factors responsible for increased supply chain costs such as excess number of intermediaries, cultivation and transportation delays due to climatic and other unavoidable conditions, commodity supply and demand scenario, trade-off between cultivate-to-order and cultivate-to-store strategy (Wu et al., 2012), precision of quality, packaging, storage, transportation and handling required, etc. The optimization performed for all the above-mentioned factors result into an efficient and responsive supply chain. The approaches utilized are nonlinear optimization for cost minimization (Pauls-worm et al., 2015; Mishra et al., 2016; Hyland et al., 2016), optimal network design (Nourbakhsh et al., 2016), dynamic pricing (Schlosser, 2015) and vehicle routing with the help of PSO (Seyedhosseini and Ghoreyshi, 2014), etc.

- *Optimal ordering policy at retail (inventory management):* The researchers in the literature have always performed the research toward optimal ordering strategies (Sarker et al., 2000; Chen and Zhou, 2014), EOQ models (Dobson et al., 2017), stochastic programming, i.e., FIFO and LIFO models

(Pauls-worm et al., 2013; Lee et al., 2014), sensitivity analysis for inventory model (Malik et al., 2016) and forecasting studies (Gao, 2015; Priyadarshi et al., 2019a) to reduce the spoilage levels at retail and assisting into effective distribution when performed as a two-stage study.

- *Boosting/enhancing demand and supply ratio at centralized markets:* The researches performed with promotions, discounts and dynamic pricing are a few examples to boost the supply and demand ratio at retail. Meanwhile, in order to increase the supply from rural levels to centralized levels further requires farmer motivation, enhanced minimum support prices (MSP), improved transportation infrastructure and inviting the food processing firms to markets to potentially develop business prospects. The simulation study with abovementioned factors have and will motivate the society, government and business agencies to move forward for economic development of farming individuals. The approaches observed for this study were stochastic programming approach for production planning (Pauls-worm et al., 2013) and demand optimization approach (Hamidi et al., 2017).

- *Information infrastructure for effective distribution of local production:* In order to cope with the existing spoilage scenario and uneven distribution of fresh commodities over the markets, there is a need for effective distribution of local production. There is a huge need for development of information infrastructure to be able to efficiently distribute the local production to the local markets, centralized markets, distant markets and exports to match the supply and demand. This could become possible with quality monitoring with cold-chain temperature management (Aung and Chang, 2014) and implementation of block-chain technology (Guo, 2018). The optimization approaches could curb this issue with presumed demand over the markets and estimated yearly yields looking at annual/seasonal rainfalls.

- *Quality parameters for food safety:* The optimal quality parameter for decision-making could include a set of criteria for fresh produce such as moisture loss, weight loss percentage allowable during acceptance, Likert-scale qualitative assessment, guidelines for cold chains, etc. Thus, the contribution of optimization and statistical studies in the field of cold chains will enhance the food quality and lesser food outbreaks such as quality degradation model (Rong et al., 2011), hybrid optimization approach for logistics network design (De Keizer et al., 2015) and nonlinear programming to optimize storage temperature to ensure quality (Mishra et al., 2016), etc.

- *Implementation of vertical integration at rural level:* The level of value addition needs to be practiced at rural level for a variety of produce to improve the social status of a farmer. The value addition process is defined as a sequence of operations performed on the produce that increases its price and quality value. Vertical integration is to perform the value addition up to an extent that the produce could be sold into markets in finished form. There is a huge need to motivate the rural population to perform value addition and vertical integration (Gao, 2015). The statistical studies, the optimization studies and empirical studies considering the produce inputs, value addition costs (i.e., fixed costs and variable costs), etc. could inspire the budding entrepreneurs to perform value addition at rural level (Priyadarshi et al., 2019b).

- *Structural Equation Modelling (SEM) on PHSC:* SEM is an advanced multivariate statistical method performed to justify the structural relationships among the set of variables with the help of hypothesized model. SEM is performed to illustrate the applicability of the hypothesized model into the real-world situation. The input data is collected in various subtypes to perform factor analysis followed by multiple regression. SEM is applied over a wide variety of real-world problems, including the supply chain issues as shown in Table 7.1. Although, finding various potential perspectives holds an infinite opportunity to be explored, such as elements of fruit consumption (Menozzi and Mora, 2012), crop recovery during post-disaster situations (Sheu, 2016), food waste behavior (Diaz-Ruiz, 2018), food losses (Amentae, 2018) and online-to-offline (O2O) food delivery services (Roh and Park, 2019), etc.

The above-mentioned research interests could enhance rural employability, farmer profitability and could reduce the distress among the farmers with lesser spoilage ratios due to availability of warehousing and effective distribution, with reduced demand and supply mismatch at markets and exports, and practice of value addition. The empirical approaches, namely, optimization, statistical and mathematical approaches are required to solve these issues with specific problem-solving perspectives.

7.4   CONCLUSIONS

The PHSC studies focusing on the losses are quantitative in nature and thus optimization and statistical methods are reported in the literature to model and minimize them. The PHSC optimization literature survey hereby indicates the directions that the researchers have worked upon and the approaches necessary for development. These approaches shall be practiced to improve rural employment scenario, social status and reduce the distress among rural populations with the adaption of various modern agricultural, handling, packaging, storage and value addition techniques. The quality parameters are to be matched by yielding quality produce and with lesser crop failures, cultivation of commercial crops, crop insurance, certification, and with practice of vertical integration alternatives (Priyadarshi and Routroy, 2018). However, it is worth mentioning that all of these methods require empirical justification to convince the government, business firms, entrepreneurs and farming individuals to take the issues in consideration. Warehousing needs to be improved to accommodate an array of produce with different shelf lives. The storage conditions during transportation, the transportation routes and number of transit points are also to be optimized to ensure produce quality. The optimal ordering at retail and sales prediction shall be performed to handle the demand optimally. The modelling approaches need to consider encouraging business ideas, technology and government policies, impact of awareness, etc.

REFERENCES

Adenso-Díaz, B., Lozano, S. and Palacio, A. 2017. Effects of dynamic pricing of perishable products on revenue and waste, *Applied Mathematical Modelling*, 45:148–164.

Ahumada, O. and Villalobos, J.R. 2009, Application of planning models in the agri-food supply chain: A review, *European Journal of Operational Research*, 196:1–20.

Ahumada, O. and Villalobos, J.R. 2011, A tactical model for planning the production and distribution of fresh produce, *Annals of Operations Research*, 190:339–358.

Ambler, K., de Brauw, A. and Godlonton, S. 2018. Measuring postharvest losses at the farm level in Malawi, *Australian Journal of Agricultural and Resource Economics*, 62:139–160.

Amentae, T.K. 2018. Supply chain management approach to reduce food losses, PhD Dissertation, Swedish University of Agricultural Sciences.

Amentae, T.K. Hamo, T.K. Gebresenbet, G. and Ljungberg, D. 2017. Exploring wheat value chain focusing on market performance, post-harvest loss, and supply chain management in Ethiopia: The case of Arsi to Finfinnee market chain, *Journal of Agricultural Science*, 9(8):22–42.

An, K. and Ouyang, Y. 2016. Robust grain supply chain design considering post-harvest loss and harvest timing equilibrium, *Transportation Research Part E: Logistics and Transportation Review*, 88:110–128.

Ansah, I.G.K., Bright, K.D.T. and Donkoh, S.A. 2017. Determinants and income effect of yam postharvest loss management: Evidence from the Zabzugu district of northern Ghana, *Food Security*, 9(3):611–620.

Ansari, M. and Smith, J.S. 2017. Warehouse operations data structure (WODS): A data structure developed for warehouse operations modeling, *Computers and Industrial Engineering*, 112:11–19.

Aung, M.M. and Chang, Y. S. 2014. Temperature management for the quality assurance of a perishable food supply chain, *Food Control*, 40:198–207.

Banasik, A. Kanellopoulos, A., Claassen, G.D.H., Bloemhof-Ruwaard, J.M. and van der Vorst, J.G.A.J. 2017. Closing loops in agricultural supply chains using multi-objective optimization: A case study of an industrial mushroom supply chain, *International Journal of Production Economics*, 183:409–420.

Barker, N., Kumar, D. and Singh, N. 2017. An economic analysis of post harvest losses of selected vegetables in Allahabad district, Uttar Pradesh, *International Journal of Current Advanced Research*, 6(2):2205–2208.

Begum, M.E.A., Hossain, Mo.I. and Papanagiotou, D.E. 2012. Economic analysis of post-harvest losses in food grains for strengthening food security in northern regions of Bangladesh, *International Journal of Applied Research in Business Administration & Economics*, 1(3):56–65.

Bendinelli, W.E., Su, C.T. and Péra, T.G. 2019. What are the main factors that determine post-harvest losses of grains? *Sustainable Production and Consumption*, 221:228–238.

Bourne, M.C. 1977. Post-harvest food lossses – The neglected dimension in increasing the world food supply, *Cornell International Agriculture Mimeograph*, 53:1–56.

Brewin, D.G. 2016. Competition in Canada's agricultural value chains: The case of grain, *Canadian Journal of Agricultural Economics*, 64:5–19.

Busch, G. and Spiller, A. 2016. Farmer share and fair distribution in food chains from a consumer's perspective, *Journal of Economic Psychology*, 55:149–158.

Chen, X. and Zhou, Q. 2014. Loss-averse retailer's optimal ordering policies for perishable products with customer returns, *Mathematical Problems in Engineering*, 2014:1–5.

Dabbene, F., Gay, P. and Sacco, N. 2008. Optimisation of fresh-food supply chains in uncertain environments, Part II: A case study, *Biosystems Engineering*, 99(3):360–371.

De Keizer, M., Haijema, R., Bloemhof, J.M. and Van Der Vorst, J.G.A.J. 2015. Hybrid optimization and simulation to design a logistics network for distributing perishable products, *Computers and Industrial Engineering*, 88:26–38.

Diaz-Ruiz, Raquel. 2018. Understanding food waste behaviours along the food supply chain – A multilevel approach. PhD Dissertation., Universitat Politècnica de Catalunya.

Dobson, G., Pinker, E.J. and Yildiz, O. 2017. An EOQ model for perishable goods with age-dependent demand rate, *European Journal of Operational Research*, 257:84–88.

East, A.R. 2011. Accelerated libraries to inform batch sale scheduling and reduce postharvest losses of seasonal fresh produce, *Biosystems Engineering*, 109:1–9.

Gao, L. 2015. Collaborative forecasting, inventory hedging and contract coordination in dynamic supply risk management, *European Journal of Operational Research*, 245:133–145.

Gardas, B.B., Raut, R.D. and Narkhede, B. 2017. Modeling causal factors of post-harvesting losses in vegetable and fruit supply chain: An Indian perspective, *Renewable and Sustainable Energy Reviews*, 80:1355–1371.

Ge, H., Nolan, J., Gray, R., Goetz, S. and Han, Y. 2016. Supply chain complexity and risk mitigation – A hybrid optimization–simulation model, *International Journal of Production Economics*, 179:228–238.

Govindan, K. Jafarian, A., Khodaverdi, R. and Devika, K. 2014. Two-echelon multiple-vehicle location-routing problem with time windows for optimization of sustainable supply chain network of perishable food, *International Journal of Production Economics*, 152:9–28.

Guan, C., Zhang, Z., Liu, S. and Gong, J. 2019. Multi-objective particle swarm optimization for multi-workshop facility layout problem, *Journal of Manufacturing Systems*, 53:32–48.

Guan, J. and Lin, G. 2016. Hybridizing variable neighborhood search with ant colony optimization for solving the single row facility layout problem, *European Journal of Operational Research*, 248(3):899–909.

Guo, X. 2018. Inventory report: Dutch smart chains for transport of perishable products, 1–54. https://research.wur.nl/en/publications/inventory-report-dutch-smart-chains-fortransport-of-perishable-pr

Hamidi, M.R., Gholamian, M.R., Shahanaghi, K. and Yavari, A. 2017. Reliable warehouse location-network design problem under intentional disruption, *Computers and Industrial Engineering*, 113:123–134.

Heenkenda, S. and Chandrakumara, D.P.S. 2016. Entrepreneurial skills and farming performance: Implications for improving banana farming in Sri Lanka, *International Journal of Humanities and Social Sciences*, 7:14–26.

Heydari, J., Mahmoodi, M. and Taleizadeh, A.A. 2016. Lead time aggregation: A three-echelon supply chain model, *Transportation Research Part E: Logistics and Transportation Review*, 89:215–233.

Hyland, M.F., Mahmassani, H.S. and Bou Mjahed, L. 2016. Analytical models of rail transportation service in the grain supply chain: Deconstructing the operational and economic advantages of shuttle train service, *Transportation Research Part E: Logistics and Transportation Review*, 93:294–315.

Ibrahim, H.I., Saba, S.S. and Ojoko, E.A. 2018. Post-harvest loss in rice production: evidence from a rural community in northern Nigeria, *FUDMA Journal of Sciences (FJS)*, 2:17–22.

Kale, A.P. and Sonavane, S. P. 2019. IoT based smart farming: Feature subset selection for optimized high- dimensional data using improved GA based approach for ELM, *Computers and Electronics in Agriculture*, 161:225–232.

Kasso, M. and Bekele, A. 2016. Post-harvest loss and quality deterioration of horticultural crops in Dire Dawa region Ethiopia, *Journal of the Saudi Society of Agricultural Sciences*, 17:88–96.

Kereth, G.A., Lyimo, M., Mbwana, H.A., Mongi, R.J. and Ruhembe, C.C. 2013. Assessment of post-harvest handling practices: Knowledge and losses of fruits in Bagamoyo district of Tanzania, *Food Science and Quality Management*, 11:8–15.

Kong, Q., Kuriyan, K., Shah, N. and Guo, M. 2019. Development of a responsive optimisation framework for decision-making in precision agriculture, *Computers & Chemical Engineering*, 131:1–33.

Laguerre, O., Hoang, H.M. and Flick, D. 2013. Experimental investigation and modelling in the food cold chain: Thermal and quality evolution, *Trends in Food Science and Technology*, 29(2):87–97.

Lamsal, K., Jones, P.C. and Thomas, B.W. 2016. Harvest logistics in agricultural systems with multiple, independent producers and no on-farm storage, *Computers and Industrial Engineering*, 91:129–138.

Lee, Y. M., Mu, S., Shen, Z. and Dessouky, M. 2014. Issuing for perishable inventory management with a minimum inventory volume constraint, *Computers and Industrial Engineering*, 76:280–291.

Liu, J., Zhang, H., He, K. and Jiang, S. 2018. Multi-objective particle swarm optimization algorithm based on objective space division for the unequal-area facility layout problem, *Expert Systems with Applications*, 102:179–192.

Majiwa, E., Lee, B.L., Wilson, C., Fujii, H. and Managi, S. 2018. A network data envelopment analysis (NDEA) model of post-harvest handling: The case of Kenya's rice processing industry, *Food Security*, 10:631–648.

Malik, A.K., Shekhar, C., Vashisth, V., Chaudhary, A.K. and Singh, S.R. 2016. Sensitivity analysis of an inventory model with non-instantaneous and time-varying deteriorating items, AIP Conference Proceedings, 1715, 020059: 1–9.

Mendoza, J.R., Sabillón, L., Martinez, W., Campabadal, C., Hallen-Adams, H.E. and Bianchini, A. 2017. Traditional maize post-harvest management practices amongst smallholder farmers in Guatemala, *Journal of Stored Products Research*, 71:14–21.

Menozzi, D. and Mora, C. 2012. Fruit consumption determinants among young adults in Italy: A case study, *LWT - Food Science and Technology*, 49(2):298–304.

Mirmajlesi, S.R. and Shafaei, R. 2016. An integrated approach to solve a robust forward/reverse supply chain for short lifetime products, *Computers and Industrial Engineering*, 97:222–239.

Mishra, A., Buchanan, R.L., Schaffner, D.W. and Pradhan, A.K. 2016. Cost, quality, and safety: A nonlinear programming approach to optimize the temperature during supply chain of leafy greens, *LWT – Food Science and Technology*, 73:412–418.

Mizgier, K.J. 2017. Global sensitivity analysis and aggregation of risk in multi-product supply chain networks, *International Journal of Production Research*, 55:130–144.

Mowrey, C.H., Parikh, P.J. and Gue, K.R. 2018. *A Model to Optimize Rack Layout in a Retail Store*, 271:1100–1112.

Nagare, M., Dutta, P. and Kambli, A. 2013. Retail inventory management for perishable products with two bins strategy, *World Academy of Science, Engineering and Technology*, 7(4):223–228.

Ndraha, N., Hsiao, H., Vlajic, J., Yang, M. and Lin, H. V. 2018. Time-temperature abuse in the food cold chain: Review of issues, challenges, and recommendations, *Food Control*, 89:12–21.

Nourbakhsh, S.M., Bai, Y., Maia, G.D.N., Ouyang, Y. and Rodriguez, L. 2016. Grain supply chain network design and logistics planning for reducing post-harvest loss, *Biosystems Engineering*, 151:105–115.

Ortmann, F.G. 2005. Modelling the South African fresh fruit export supply chain, Master's Thesis, University of Stellenbosch, South Africa, 1–292.

Pauls-worm, K.G.J., Hendrix, E.M.T., Haijema, R. and Der, van der Vorst, J.G.A.J. 2013. Inventory control for a perishable product with non-stationary demand and service level constraints, *Operations Research and Logistics*, Wageningen: optimization-online.org, 2013:1–22. www.optimization-online.org/DB_HTML/2013/08/4010.html

Pauls-worm, K. G. J., Hendrix, E. M. T., Alcoba, A. G. and Haijema, R. 2015. Order quantities for perishable inventory control with non-stationary demand and a fill rate constraint, *International Journal of Production Economics*, 181:1–9.

Priyadarshi, R., Panigrahi, A., Routroy, S. and Kant, Girish. 2019a. Demand forecasting at retail stage for selected vegetables: A performance analysis, *Journal of Modelling in Management*, 14(4):1042–1063.

Priyadarshi R. and Routroy, S. 2018. Vertical integration level selection for value addition of herbal products: A farmer's perspective, *Materials Today: Proceeding*, 5(9):18354–18361.

Priyadarshi, R., Routroy S. and Kant, Girish. 2019b. Analysis of enablers for vertical integration to enhance rural employability, *Journal of Business and Industrial Marketing*, 34(4):690–702.

Rabbani, M., Farshbaf-Geranmayeh, A. and Haghjoo, N. 2016. Vehicle routing problem with considering multi-middle depots for perishable food delivery, *Uncertain Supply Chain Management*, 4:171–182.

Raut, R.D., Gardas, B.B., Kharat, M. and Narkhede, B. 2018. Modeling the drivers of post-harvest losses – MCDM approach, *Computers and Electronics in Agriculture*, 154:426–433.

Roh, M. and Park, K. 2019. Adoption of O2O food delivery services in South Korea: The moderating role of moral obligation in meal preparation, *International Journal of Information Management*, 47:262–273.

Rong, A., Akkerman, R. and Grunow, M. 2011. An optimization approach for managing fresh food quality throughout the supply chain, *International Journal of Production Economics*, 131:421–429.

Sarker, B.R., Jamal, A.M.M. and Wang, S. 2000. Supply chain models for perishable products under inflation and permissible delay in payment, *Computers & Operations Research*, 27:59–75.

Schlosser, R. 2015. Dynamic pricing and advertising of perishable products with inventory holding costs, *Journal of Economic Dynamics and Control*, 57, 163–181.

Seyedhosseini, S.M. and Ghoreyshi, S.M. 2014. An integrated model for production and distribution planning of perishable products with inventory and routing considerations, *Mathematical Problems in Engineering*, 2014:1–11.

Shekhawat, K. 2018. Enumerating generic rectangular floor plans, *Automation in Construction*, 92:151–165.

Shekhawat, K. 2015. Why golden rectangle is used so often by architects: a mathematical approach, *Alexandria Engineering Journal*, 54:213–222.

Sheu, J.B. 2016. Supplier hoarding, government intervention, and timing for post-disaster crop supply chain recovery, *Transportation Research Part E: Logistics and Transportation Review*, 90:134–160.

Song, B.D. and Ko, Y.D. 2016. A vehicle routing problem of both refrigerated- and general-type vehicles for perishable food products delivery, *Journal of Food Engineering*, 169:61–71.

Springael, J., Paternoster, A. and Braet, J. 2018. Reducing postharvest losses of apples: Optimal transport routing (while minimizing total costs), *Computers and Electronics in Agriculture*, 146:136–144.

Sternbeck, M.G. and Kuhn, H. 2014. An integrative approach to determine store delivery patterns in grocery retailing, *Transportation Research Part E: Logistics and Transportation Review*, 70:205–224.

Tinello, D., Jodin, D. and Winkler, H. 2016. Biomimetics applied to factory layout planning: Fibonacci based patterns, spider webs and nautilus shell as bio-inspiration to reduce internal transport costs in factories, *CIRP Journal of Manufacturing Science and Technology*, 13:51–71.

Trebuna, P., Pekarcikova, M. and Edl, M. (2019). Digital value stream mapping using the tecnomatix plant simulation software. *International Journal of Simulation Modelling*, 18:18–32.

Wang, X., Wang, M., Ruan, J. and Zhan, H. 2016. The multi-objective optimization for perishable food distribution route considering temporal-spatial distance, *Procedia Computer Science*, 96:1211–1220.

Wu, Y., Cheng, T.C.E. and Zhang, J. 2012. A serial mixed produce-to-order and produce-in-advance inventory model with multiple retailers, *International Journal of Production Economics*, 136(2):378–383.

Xiao, R., Xu, Y. and Amos, M. 2007. *Two Hybrid Compaction Algorithms for the Layout Optimization Problem*, 90:560–567.

Zhao, T.S., Wu, K. and Yuan, X. 2016. Optimal production-inventory policy for an integrated multi-stage supply chain with time-varying demand, *European Journal of Operational Research*, 255(2):364–379.

WEBSITES

Wikipedia. (2019). Indian railways. https://en.wikipedia.org/wiki/Indian_Railways (Accessed on 15 April 2020).

Economic Times. (2019). "Air freight demand may touch 1.1 mt by 2025: Report". https://economictimes.indiatimes.com/industry/transportation/airlines-/-aviation/air-freight-demand-may-touch-1-1-mt-by-2025-report/articleshow/69775888.cms?from=mdr (Accessed on 15 April 2020).

Indiastat (2019a). https://www.indiastat.com/table/agriculture-data/2/warehouses/206862/394668/data.aspx (Accessed on 15 April 2020).

Indiastat (2019b). https://www.indiastat.com/table/agriculture-data/2/warehouses/206862/1200234/data.aspx (Accessed on 15 April 2020).

CNKI. 2015. http://en.cnki.com.cn/Article_en/CJFDTotal-SJSJ201504012.htm (Accessed on 15 April 2020).

walkthroughindia.com (2019), http://www.walkthroughindia.com/walkthroughs/11-most-amazing-facts-about-the-indian-road-network/#:~:text=The%20Road%20network%20of%20India,major%20district%20and%20rural%20roads (Accessed on 15 April 2020).

8 Prioritization of Barriers for E-Waste Management in India Using Best Worst Method

Shivam Goyal, Vernika Agarwal
Amity International Business School, Amity University
Noida, India

CONTENTS

8.1   INTRODUCTION

The growth of e-waste in developing countries is putting a lot of pressure on the prevailing environment. The flow of e-waste causes a threat to human lives; therefore, there is a need of proper waste of electrical and electronic equipment (WEEE) management. To counter this problem, the government is pressurizing the industrialists to implement proper and effective management to dispose of e-waste. This problem of WEEE prevails in mostly developing countries where a huge quantity of e-waste is imported from developed countries because there is high labor cost and strict government norms. With the increasing knowledge about e-waste, the community is becoming aware and the government is implementing necessary policies to overcome the e-waste problem. India is a primary dumping site for WEEE from the developed nations (Gaziulusoy, 2015; Govindan et al., 2019). The government has introduced and implemented a set of various policies in Extended Producer Responsibility (2011). The EPR policies were amended in 2018. The main reason for the failure of the same is lack of implementation and follow up. The reason behind it is the delay in law enforcement on producers or manufacturers, so as to come up with proper management for handling returns and proper disposal of electronic products (Dwivedy and Mittal, 2010). Lack of policies and regulations is a cause for the same. Today about 85 percent of e-waste is recycled by the unorganized sector, which uses traditional ways to recycle. The recycle process opted by them is unfriendly to nature and does a lot harm to animals, humans and plants (Nnorom et al., 2009). As the report of the associated chamber of commerce (ASSOCHAM) states, India is an emerging country that stands fifth position among the developing countries around the world and second position in Asia, and it produces e-waste with a growth of 25 percent a year, which is 18.5 lakh MT of electronic waste by 2016 as compared to the current level of 12.5 lakh MT annually (Ganguly, 2019). Therefore, with this motivation, the present work is to study the challenges that hinder the implementation of e-waste management in an Indian scenario. The research objectives for the study are as follows:

1. Identification of barriers that hinder the implementation of e-waste management in India.
2. Analysis and identification of the influential barrier with the help of BWM approach through experts' opinions.

To achieve these objectives, this chapter presents a framework for evaluating the normal and governmental barriers that hinder the proper implementation of e-waste management to aid the Indian industries. Initially, a literature survey is carried out to identify the potential normal and governmental barriers. Best worst method (BWM) is used to analyze the shortlisted attributes, determine the weights (relative importance) of these barriers and finally prioritize them.

The novelty of the study, which distinguishes it from the other studies in this field, lies in identifying the normal and governmental barriers and application of BWM. There are many studies in the literature that identify the barriers for e-waste management; however, there is lack of studies that identify governmental as well as normal barriers. Kumar and Dixit (2018) identified ten barriers, mostly governmental in nature, and used interpretive structural modeling (ISM) and Decision Making Trail and Evaluation Laboratory (DEMATEL) for understanding the hierarchal and contextual relationship structure among them. They concluded that the lack of public awareness about e-waste recycling and the lack of policies addressing e-waste issues are the root cause barriers in Indian context. Bhatia and Srivastava (2018) used a grey-DEMATEL approach to analyze the interrelationship among barriers for e-waste management in India. Satapathy et al. (2018) ranked the barriers for e-waste management in India based on their importance by using Preference Ranking Organization Method for Enrichment Evaluation (PROMETHEE II) and VlseKriterijumska Optimizacija I Kompromisno Resenje (VIKOR) analysis. Chauhan et al. (2018) focused on governmental barriers by using an ISM-DEMATEL approach. The findings of the study suggest that the lack of funds, input material and subsidy are the most influential barriers that are needed to be addressed in an Indian context. Pires et al. (2019) analyzed the major challenges to waste collection and management in developing and developed countries. The focus of the work was to better understand the growing informal sector in e-waste management. The focus of the literature is to analyze the governmental barriers. While the literature does not particularly discuss the incorporation of normal as well as governmental barriers, this chapter addresses these gaps and focuses on developing a framework for analyzing normal and governmental barriers using BWM. The advantage of using BWM is that it requires fewer comparison matrices as compared to any multicriteria technique. Hence, the basic objective of this study is twofold. First, we aim to identify the major barriers that hinder the proper implementation of e-waste management. In the second phase, a relatively newer technique known as BWM is used to evaluate the weights and ranks of different barriers. The organization of the manuscript is as follows. Section 8.2 presents the research methodology of the BWM methodology. Section 8.3 represents the numerical illustration to validate the proposed model. Section 8.4 addresses the conclusion of the chapter and a brief description of future work.

8.2 RESEARCH METHODOLOGY

This section elaborates the research methodology. We have used a relatively newer methodology, BWM, to calculate the weights of the various governmental and normal barriers that hinder the proper management of e-waste in India. The perspectives of the key decision makers (DMs) are used for the overall assessment of the system. In the first phase, we have identified the various governmental and normal barriers using literature survey and interactions with the DMs. In the second phase, we have used BWM to calculate the weights of these barriers. BWM was developed by Jafar Rezaei in 2015 and is a relatively newer approach that conducts structured pair-wise comparisons with relatively lesser information as compared to other Multi Criteria Decision Making (MCDM) techniques (Rezaei, 2015).

8.2.1 IDENTIFICATION OF THE BARRIERS

In this phase, the various barriers are identified based on literature survey and discussion with the stakeholders.

8.2.2 PRIORITIZATION OF THE BARRIERS

The research methodology used in the same is BWM. The steps for the same are:

- **Select the most and least prominent barriers.**
  In this step we have analyzed the most prominent and least prominent barriers that hinder the implementation of e-waste management.
- **Compare barriers.**
  In this step we have compared the most prominent barrier with the others and the least prominent barrier with the other barriers. The score used is 1–9.
  The resulting vector of "Best-to-Others" is given as:

$$SV_B = \left(sv_{B1},...,sv_{B9}\right)$$

where sv_{Bi} gives the preference of the most prominent barrier over others and $sv_{BB} = 1$.

The resulting vector of "Worst-to-Others" is given as:

$$SV_W = \left(sv_{W1},...,sv_{W9}\right)$$

where sv_{Wi} gives the preference of the other barriers over least prominent barriers and $sv_{WW} = 1$.

- **Calculate optimal weights for the barriers.**
  The aim of this step is to calculate the optimal weighting vector $\left(z_1^*,...,z_9^*\right)$ of the barriers.

  The optimal weight of ith barrier is the one which meets the following requirements:

$$\frac{z_B^*}{z_i^*} = sv_{Bi} \quad \text{and} \quad \frac{z_i^*}{z_W^*} = sv_{iw}.$$

In order to satisfy this condition, the maximum absolute difference $\left|\dfrac{z_B^*}{z_i^*} - sv_{Bi}\right|$ and $\left|\dfrac{z_i^*}{z_W^*} - sv_{iw}\right|$ need to be minimized for all problem.

Hence, the optimal weighs can be achieved through the following programming problem:

$$\min_i \max \left\{ \left|\frac{z_B^*}{z_i^*} - sv_{Bi}\right| , \left|\frac{z_i^*}{z_W^*} - sv_{iw}\right| \right\}$$

Subject to

$$\sum_I z_i = 1 \tag{P1}$$

$$z_i \geq 0 \qquad \forall i = 1,2,...,n;$$

Problem (P1) is equivalent to the following linear programming formulation (P2):

$$\min \phi$$

Subject to

$$\left|z_B - sv_{Bi}z_i\right| \leq \phi \qquad \forall i = 1,2,...,n;$$
$$\left|z_i - sv_W z_W\right| \leq \phi \qquad \forall i = 1,2,...,n;$$
$$\sum_I z_i = 1 \tag{P2}$$
$$z_i \geq 0 \qquad \forall i = 1,2,...,n;$$

TABLE 8.1
Consistency Index Table for BWM

v_{Bi}	1	2	3	4	5	6	7	8	9
Consistency index (max) 0.00	0.44	1.00	1.63	2.30	3.00	3.73	4.47	5.2	

On solving the above problem (P2), we get the value of consistency ratio ϕ^* and optimal weights as $\left(z_1^*,...,z_n^*\right)$. The closer the consistency ratio ϕ^* is to zero value, the more consistent the comparison system.

• **Check the consistency of the solution.**
The closer the consistency ratio is to zero value, the more consistent the comparison system provided by the DM. We check the consistency of the solution by calculating the consistency ratio:

$$\text{Consistency Ratio} = \frac{\phi^*}{\text{Consistency Index}}$$

Table 8.1 is used to get the value of the consistency index (Rezaei, 2015).

The value of consistency ratio closer to 0 shows more consistency, while values closer to 1 show less consistency.

8.3 NUMERICAL ILLUSTRATION

Application of BWM is focused in two phases, such as research design and data collection. The details of each phase were summarized in sections 3.1 and 3.2, which will be briefly discussed in the following sections.

8.3.1 RESEARCH DESIGN

The aim of this research design is to identify the major barriers that are being faced for proper management of e-waste. The barriers used in this manuscript have been taken from literature survey as well as interactions with various stakeholders. The present study targeted NGO workers, e-waste recyclers and various electronic retailers from National Capital Region, Delhi, India to identify the prominent barriers. An expert panel was formed consisting of two NGO workers, three recyclers and three retailers to discuss and finalize the barriers. The identified barriers consist of seven types of governmental barriers as well as six types of normal barriers that have been identified by from National Capital Region (India). Tables 8.2 and 8.3 give the shortlisted governmental and normal barriers.

8.3.2 APPLICATION OF THE BEST WORST METHODOLOGY (BWM) FOR THE CASE

To elaborate on the evaluation process, we have discussed a step-by-step evaluation process in this section. Once we have finalized the barriers, the relevant data was

TABLE 8.2
List of Governmental Barriers

Notation	Barriers (GOVERNMENTAL)	References
GR1	There is lack of proper policy that considers all aspects related to e-waste management, including assignment of responsibilities for all stakeholders	Chowrimootoo (2011)
GR2	The policies sometimes lack practical and rational issues, and so the sustainability issue is perhaps not well addressed	Great Lakes Electronics Corporation (2019)
GR3	Competition between formal and informal sectors	Chowrimootoo (2011)
GR4	There is a lack of standard infrastructure in developing to deal with ICT waste	Ryder (2019)
GR5	Lack of technology and skills for the industries involved in the treatment	Chowrimootoo (2011)
GR6	Lack of economic support and consultancy to know where to invest to maximize the value added and obviously improve and maintain the effectiveness of the treatment industry	Great Lakes Electronics Corporation (2019)
GR7	Including limited access to recycling bin containers, unreliable collection service or living in places far from recycling sites	Ryder (2019)

TABLE 8.3
List of Normal Barriers

Notation	Barriers (NORMAL)	References
NB1	Lack of reliable information regarding the amount and categories of e-waste to be treated, which eventually makes it difficult to devise the correct strategy to be used and ultimately to invest correctly in treatment industries	Great Lakes Electronics Corporation (2019)
NB2	Lack of correct standards and procedures for the collection and eventual treatment of e-waste	Ryder (2019)
NB3	Low recycling penetration and low supply of domestic e-waste	OWN
NB4	Attitudes or perceptions, such as thinking recycling is a waste of time	Great Lakes Electronics Corporation (2019)
NB5	Behavioral issues, like people assuming they're too busy to recycle or simply forget about it	Ryder (2019)
NB6	Lack of information on reuse and recycling possibilities	OWN

TABLE 8.4

Best-to-Others and Others-to-Worst Vectors for First Stakeholder

Best-to-Others	NB1	NB2	NB3	NB4	NB5	NB6
NB1	1	3	4	2	6	5
Others-to-Worst	NB5					
NB1	6					
NB2	4					
NB3	3					
NB4	5					
NB5	1					
NB6	2					

collected using the opinion of the expert panel. The expert panel used for collection of data include three recyclers. We asked the three recyclers to select the prominent and least prominent barriers and give reference comparison values from a 1–9 scale. This data was used to generate weights using the BWM technique as elaborated in section 3.2. This was done separately for government and normal barriers so as to understand the prominent barriers in each category. For the first stakeholder, let the most prominent normal barrier be NB1– "Lack of reliable information regarding the amount and categories of e-waste to be treated, which eventually makes it difficult to devise the correct strategy to be used and ultimately to invest correctly in treatment industries". The least prominent barrier is NB5 –"Behavioural issues: like people assuming they're too busy to recycle or simply forget about it". The next step is to quantify the best-to-others and others-to-worst vectors, which are given in Table 8.4.

On solving this, using the steps of the BWM methodology, we get the following weights NB1 – 0.379, NB2 – 0.148, NB3 –0.110, NB4 – 0.221, NB5 – 0.53 and NB6 – 0.089. Figure 8.1 gives the graphical representation of these weights.

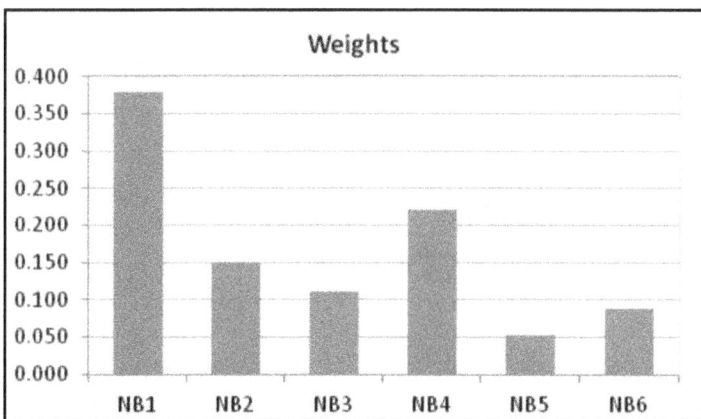

FIGURE 8.1   Graphical representation of optimal weights.

TABLE 8.5

Best-to-Others and Others-to-Worst Vectors for Government Barriers

Government Barriers	Best-to-Others	Others-to-Worst
GR1	2, 2, 1	5, 5, 7
GR2	5, 6, 5	2, 5, 3
GR3	6, 7, 7	1, 1, 1
GR4	1, 5, 4	6, 2, 4
GR5	4, 3, 2	3, 6, 6
GR6	2, 1, 6	5, 7, 2
GR7	3, 4, 3	4, 3, 5

Similarly, we collect the responses of all the stakeholders for the most prominent and least prominent governmental and normal barriers. Tables 8.5 and 8.6 provide the responses of the stakeholders.

It can be seen that different stakeholders have given different ranking to each of the barriers based on the difficulty faced by them in proper implementation of the e-waste management. In order to find the optimal weight, we follow the steps of the BWM method. The optimal weights are given in Tables 8.6 and 8.7.

TABLE 8.6

Best-to-Others and Others-to-Worst Vectors for Normal Barriers

Normal Barriers	Best-to-Others	Others-to-Worst
NB1	1, 2, 1	6, 5, 8
NB2	3, 5, 5	4, 2, 5
NB3	4, 3, 2	3, 4, 6
NB4	2, 1, 3	5, 6, 3
NB5	6, 6, 6	1, 1, 2
NB6	5, 4, 8	2, 3, 1

TABLE 8.7

Optimal Weights of Governmental Barriers

Normal Barrier	GR1	GR2	GR3	GR4	GR5	GR6	GR7	Optimal Weight
Stakeholder 1	0.353	0.084	0.041	0.105	0.209	0.070	0.139	0.066
Stakeholder 2	0.217	0.072	0.034	0.087	0.145	0.336	0.109	0.098
Stakeholder 3	0.181	0.072	0.043	0.311	0.091	0.181	0.121	0.052
Average weight	0.250	0.076	0.039	0.167	0.148	0.196	0.123	0.072

TABLE 8.8
Optimal Weights of Normal Barriers

Normal Barrier	NB1	NB2	NB3	NB4	NB5	NB6	Optimal Weight
Stakeholder 1	0.379	0.148	0.111	0.221	0.053	0.089	0.063
Stakeholder 2	0.388	0.096	0.239	0.160	0.080	0.037	0.090
Stakeholder 3	0.221	0.089	0.148	0.379	0.053	0.111	0.063
Average weight	0.330	0.111	0.166	0.253	0.062	0.079	0.072

It can be seen that the overall weights are consistent in nature, thus we can accept this ranking.

Proper management of e-waste is a major concern for developing economies like India. With the help of the BWM approach, we have analyzed the most and least prominent barrier that hinders e-waste management. It was found that the most prominent governmental barrier was GR1 – "There is lack of proper policy that considers all aspects related to e-waste management including assignment of responsibilities for all stakeholders"; while normal barrier comes out to be NB1 – "Lack of reliable information regarding the amount and categories of e-waste to be treated, which eventually makes it difficult to devise the correct strategy to be used and ultimately to invest correctly in treatment industries".

8.4 CONCLUSION

With the rapid growth of electronic sector, the problem of e-waste is becoming a major concern for the government. Although the government is undertaking a number of schemes to encounter these problems, there is still a huge gap in the implementation of the same. This is specifically true for mostly developing countries where, in spite of strict government norms, India is still a primary dumping site for WEEE. In this direction, the primary requirement is to understand the major factors that hinder the proper management of e-waste. The identified barriers consist of seven types of governmental barriers as well as six types of normal barriers that have been identified by from National Capital Region (India). This study used a recently developed methodology of BWM for identifying the prominent barrier based on the opinion of the expert panel. The opinions of the various recyclers are considered for validating the study. The most prominent governmental barrier was found to be GR1 – "There is lack of proper policy that considers all aspects related to e-waste management including assignment of responsibilities for all stakeholders"; while normal barrier comes out to be NB1 – "Lack of reliable information regarding the amount and categories of e-waste to be treated, which eventually makes it difficult to devise the correct strategy to be used and ultimately to invest correctly in treatment industries". Hence, this study provides a framework for the assessment of various barriers that hinder the proper implementation of e-waste management in India.

8.4.1 Limitations and Future Scope

There are a few limitations of this study. Presently, we are using BWM for the prioritization of the barriers. With crisp inputs, we can also incorporate subjectivity into the decision-making process by the used of Fuzzy BWM. A comparative study can also be done between various MCDM techniques like VlseKriterijumska Optimizacija I Kompromisno Resenje (VIKOR), Analytic Hierarchy Process (AHP), and Elimination Et Choice Translating Reality (ELECTRE) for this study.

REFERENCES

Bhatia, M.S. and Srivastava, R.K. 2018. Analysis of external barriers to remanufacturing using grey-DEMATEL approach: An Indian perspective. *Resources, Conservation and Recycling*. 136:79–87.

Chauhan, A., Singh, A. and Jharkharia, S. 2018. An ISM and DEMATEL method approach for the analysis of barriers of waste recycling in India. *Journal of the Air and Waste Management Association*, 68(2):100–110.

Chowrimootoo, D.J.M. 2011. E-waste: Causes, hazards, barriers and approaches to effective management. https://www.academia.edu/8958150/E-waste_Causes_hazards_barriers_and_approaches_to_effective_management?auto=download.

Dwivedy, M. and Mittal, R.K. 2010. Estimation of future outflows of e-waste in India. *Waste Management*. 30(3):483–491.

Ganguly, R. 2019. Aspects of e-waste management in India. In Electronic Waste Pollution. Champaign, IL: Springer, 253–265.

Gaziulusoy, A.I. 2015. A critical review of approaches available for design and innovation teams through the perspective of sustainability science and system innovation theories. *Journal of Cleaner Production*. 107:366–377.

Govindan, K., Jha, P.C., Agarwal, V. and Darbari, J.D. 2019. Environmental management partner selection for reverse supply chain collaboration: A sustainable approach. *Journal of Environmental Management*. 236:784–797.

Great Lakes Electronics Corporation. 2019. The top barriers to e-waste recycling and how to solve them. Great Lakes Electronics Corporation (blog). Accessed 9 December 2019, https://www.ewaste1.com/top-barriers-to-e-waste-recycling/.

Kumar, A. and Dixit, G. 2018. An analysis of barriers affecting the implementation of e-waste management practices in India: A novel ISM-DEMATEL approach. *Sustainable Production and Consumption*. 14:36–52.

Nnorom, I.C., Ohakwe, J. and Osibanjo, O. 2009. Survey of willingness of residents to participate in electronic waste recycling in Nigeria–A case study of mobile phone recycling. *Journal of Cleaner Production*. 17(18):1629–1637.

Pires, A., Martinho, G., Rodrigues, S. and Gomes, M.I. 2019. Technical barriers and socioeconomic challenges. In Sustainable Solid Waste Collection and Management. Champaign, IL: Springer, 335–348.

Rezaei, J. 2015. Best-worst multi-criteria decision-making method. *Omega*. 53:49–57.

Ryder, G. 2019. The world's e-waste is a huge problem. It's also a golden opportunity. World Economic Forum (website). Accessed 9 December 2019, https://www.weforum.org/agenda/2019/01/how-a-circular-approach-can-turn-e-waste-into-a-golden-opportunity/.

Satapathy, S., Garanaik, A. and Kumar, S. 2018. Prioritising the barriers of waste management as per Indian perspective by PROMETHEE II and VIKOR methods. *International Journal of Services and Operations Management*. 29(4):462–486.

Section II

Soft Computing Models

9 Fuzzy Analysis of Batch Arrival Priority Queueing System with Balking

Anamika Jain, Sonali Thakur, Bhoopendra Pachauri
Manipal University Jaipur, Jaipur, India

Madhu Jain
Indian Institute of Technology Roorkee, Uttarakhand, India

CONTENTS

9.1   INTRODUCTION

Unreliable server queueing models are the topic of interest for the queue theorists, due to their hypothetical structure and importance, to deal with many physical congestion problems. In the present study, we analyze a priority queue with single-server service facility wherein the server is subject to random breakdowns and repairs. In several congestion situations, the discouragement behavior of nonprioritized customers may be noticed. In queueing literature, some authors explored the analytical aspects of queueing models with bulk arrivals via different techniques. Prioritized queues with batch arrivals occur in several congestion problems, predominantly in the circumstances where priority is given to some individuals; for example, we can notice the ordinary vs. priority customers in queueing scenarios for economy-class vs. business-class passengers, ordinary vs. emergency patients, data vs. voice transmission, etc. The priority queueing models can be developed for the formulation and performance prediction of delay and blocking situations to investigate the queueing characteristics of manufacturing and assembly lines, communication and computer systems and many more.

In the last few decades, queues with the customers' behavior have been studied in different frameworks and can be fitted well to the queueing systems with bulk input.

Drekic and Woolford (2005) analyzed the queueing model with customer's balking behavior in preemptive priority queue. Al-Seedy et al. (2009) proposed Markovian model with multiserver in the transient form to analyze the discouragement behavior of the customers. Choudhury et al. (2010) discussed the two-phase service model for the unreliable server queue with retrial attempts and batch arrivals. Jain and Bhagat (2013) suggested recovery policy for the server based on threshold level for an unreliable server queue with double orbits by considering the retrial attempts and priority customers. Hassan and Hoda Ibrahim (2013) analyzed multilevel queueing systems with unreliable server using recursive solution technique. The concept of mixed priority with negative arrivals was considered by Dimitriou (2013a, 2013b) while developing the model for the batch arrival priority queues with hybrid failure recovery discipline. Vadivu et al. (2014) worked on a multiserver unreliable priority queueing model with loss customers and general service time distribution for VoIP system. Dudin et al. (2015) discussed the priority-based Markovian queueing system operating in random environment. They considered that the servers are reserved when the environment changes occur in its state. Jain and Jain (2014, 2019) considered the prioritized queueing models with service interruption and bulk arrival. Ayyappan and Somasundaram (2019) suggested the priority queueing system with unreliable server under the criteria that the customers may join the retrial orbit when the ordinary service interruptions occur. Singh et al. (2020) analyzed the unreliable server, bulk arrival and negative customers retrial queueing system, and they also used the uncertainty by maximum entropy principle.

In most of the queueing literature, crisp values of parameters are considered. There may be uncertainty in the parameters as such queueing models with fuzzy parameters provide more robust results. A few researchers have done works on queueing models in fuzzy environment. Bellman and Zadeh (1970) discussed the concept of fuzziness and handled multistage decision processes. Prade (1980), Li and Lee (1989), Buckley (1990), Buckley et al. (2001) and Kao and Chen (1999) proposed the probabilistic model and emphasized the framework of fuzzy service rule in queueing problems. Ke and Lin (2006) considered the unreliable server system while analyzing the fuzzy queues. Ke et al. (2007) studied the reliability characteristics in fuzzy set up of a machine repair system (MRS) by considering the fuzzified failure and repair rates. By considering the fuzzy parameters, Chandra Shekhar et al. (2014) discussed the performance of MRS operating in the environment of switching failure and reboot delay. They have presented the set of parametric nonlinear programs to study the impact of fuzzified failure and repair times on the performance indices. Sanga and Jain (2019), Jain and Ahuja (2020) analyzed the prioritized customers in double orbit queue with balking behavior and applied the parametric nonlinear programing approach to determine fuzzified indices. Jain and Sanga (2020) considered the machine repair problem with general retrial and provided the results for fuzzified model to study the cost optimization.

In broad-spectrum, the jobs may arrive to a service station in batches. The workstation usually performs various kinds of jobs simultaneously or consecutively. In the present study, $M_1^{X_1} M_2^{X_2}/G_1 G_2/1$ queueing model with the priority customers and unreliable server is investigated. The crisp and fuzzified results of the concerned model have been compared. The model can be applicable in analyzing the priority-based queueing characteristics of various systems, such as in distributed computing system, tel-communication, etc. In section 9.2, we outline the assumptions, notations that are required for the mathematical formulation of the model. The analytical solution using supplementary variable technique (SVT) and distribution of the queue size are provided in section 9.3. The waiting time distributions of the prioritized and nonpriority customers are presented in section 9.4. By setting the appropriate parameter values, special cases are elaborated in section 9.5. In section 9.6, the numerical results and sensitivity analysis are presented. The final concluding remarks are also given in section 9.7.

9.2 MODEL DESCRIPTION

Consider a $M_1^{X_1} M_2^{X_2}/G_1 G_2/1$ model to investigate the unreliable server queueing system with prioritized customers. In the existing model, we consider two different categories of customers. The category 1 customers are called priority customers, whereas category 2 customers are called nonprioritized customers. The nonpreemptive priority is given to the customers of category 1 over the customers of category 2. The category 1 customer is always served before a category 2 customers while waiting in the queue. However, if a category 1 customer arrives and finds any customer of category 2 in service, the server cannot preempt the undergoing service, and the customer of category 1 is taken for service after the service of ongoing category 2 customer is completed – i.e., the service of the category 1 customer activates only on the completion of service of the category 2 customer. The category 2 customers may balk (lost) only if the server is busy in serving the category 1 customers.

In the identical category, the customers are served following the FCFS rule. The priority queued customer is selected first to be served as the service of ongoing customer finishes; otherwise if the prioritized customer is not present, the nonpriority customers are served in FCFS manner. As soon as the server fails for any cause, it is immediately sent for the repair to get recovery. The category 1 and category 2 customers arrive in the system in accordance with Poisson process with mean rates λ_1 and λ_2, respectively. We assume that due to baking behavior, the joining probability of the nonpriority customers in the queue is "θ". Both categories of customers arrive in batches with the batch sizes U_1 and U_2, respectively, and having distribution functions:

$$P\ (U_1 = i) = g_1(i) \text{ and } Q\ (U_2 = i) = g_2(i), \ i = 1, 2, 3, \ldots$$

The service times $B_1(.)$ and $B_2(.)$ are i.i.d. with general distribution function for the category 1 and category 2 customers. The service time probability generating functions (p.g.f.s) $b_1(.)$ for categories 1 customers and $b_2(.)$ for categories 2 customers are defined by:

$$b_1(\hat{u}) = \mu_1(\hat{u}) \exp\left\{-\int_0^{\hat{u}} \mu_1(\hat{\tau})\, d\hat{\tau}\right\}$$

$$b_2(\hat{u}) = \mu_2(\hat{u}) \exp\left\{-\int_0^{\hat{u}} \mu_2(\hat{\tau})\, d\hat{\tau}\right\}$$

Here $\beta_i(\hat{v})$ and $\mu_i(\hat{u})$, $(i = 1, 2)$ denote the repair rate and the service rate for categories 1 and categories 2 customers, respectively, and are given as follows:

$$\beta_i(\hat{v}) = \frac{r_i(\hat{v})}{1 - R_i(\hat{v})}, \quad \mu_i(\hat{u}) = \frac{b_i(\hat{u})}{1 - B_i(\hat{u})}$$

The corresponding p.g.f.s are denoted by $\bar{G}_1(\hat{z}_1)$ and $\bar{G}_2(\hat{z}_2)$, with respective means g_1' and g_2'. The server fails in Poisson fashion with rate α_1 and α_2 when rendering service to the category 1 and category 2 customers, respectively. The times required to repair the broken-down server are random variables V_1 for priority customer and V_2 for nonpriority customers with general distributions $R_1(.)$ and $R_2(.)$, respectively. The p.d.f. of repair time $r_1(.)$ for category 1 (priority) customers and $r_2(.)$ for category 2 (nonpriority) customers are defined by:

$$r_1(\hat{v}) = \beta_1(\hat{v}) \exp\left\{-\int_0^{\hat{v}} \beta_1(\hat{\tau}) \, d\hat{\tau}\right\}$$

$$r_2(\hat{v}) = \beta_2(\hat{v}) \exp\left\{-\int_0^{\hat{v}} \beta_2(\hat{\tau}) \, d\hat{\tau}\right\}$$

The service time distribution has finite l^{th} moments for category 1 and category 2 customers, given by

$$\tau_1^l = (-1)^l b_1^{*(l)} \quad \text{and} \quad \tau_2^l = (-1)^l b_2^{*(l)}$$

The lth moments of repair time distribution for category 1 and category 2 customers, respectively, are

$$\gamma_1^l = (-1)^l r_1^{*(l)} \quad \text{and} \quad \gamma_2^l = (-1)^l r_2^{*(l)}$$

9.3   THE ANALYSIS AND QUEUE SIZE DISTRIBUTION

For analysis purpose, the supplementary variable technique (SVT) is used by introducing the additional variable, corresponding to elapsed service times and elapsed repair times. The stochastic process $[N_1(\hat{t}), N_2(\hat{t}), U(\hat{t}), V(\hat{t}), S(\hat{t}), \hat{t} \geq 0]$ is a Markov process where $N_1(\hat{t})$ denote the number of category 1 (priority) customers and $N_2(\hat{t})$ denote the number of category 2 (nonpriority) customers in the system, $U(\hat{t})$ and $V(\hat{t})$ represent the elapsed service time and elapsed repair time, respectively, and $S(\hat{t})$ denotes the server's state at time t given by

$$S(\hat{t}) = \begin{cases} 0, \textit{ state of the idle server} \\ 1, \textit{ state of the busy server} \\ 2, \textit{ state of the brokendown server} \end{cases}$$

We now define the combined probabilities below:

The priority customer is in service.

$$P_{m,n,1}\left(\hat{u},\hat{\tau}\right)d\hat{u} = \Pr\{N_1\left(\hat{\tau}\right)=m,N_2\left(\hat{\tau}\right)=n,S\left(\hat{\tau}\right)=1,\hat{u}<U\left(\hat{\tau}\right)\leq\hat{u}+d\hat{u}\},m\geq 1,n\geq 0.$$

The server is under repair.

$$P_{m,n,2}\left(\hat{u},\hat{v},\hat{\tau}\right)d\hat{v} = \Pr\{N_1\left(\hat{\tau}\right)=m,N_2\left(\hat{\tau}\right)=n,S\left(\hat{\tau}\right)=2U\left(\hat{\tau}\right)=\hat{u},$$
$$\hat{v}<V\left(\hat{\tau}\right)\leq\hat{v}+d\hat{v}\},m\geq 1,n\geq 0,\hat{u}\geq 0,\hat{v}\geq 0.$$

The nonpriority customer is in service.

$$Q_{m,n,1}\left(\hat{u},\hat{\tau}\right)d\hat{u} = \Pr\{N_1\left(\hat{\tau}\right)=m,N_2\left(\hat{\tau}\right)=n,S\left(\hat{\tau}\right)=1,\hat{u}<U\left(\hat{\tau}\right)\leq\hat{u}+d\hat{u}\},m\geq 0,n\geq 1.$$

The server is under repair.

$$Q_{m,n,2}\left(\hat{u},\hat{v},\hat{\tau}\right)d\hat{v} = \Pr\{N_1\left(\hat{\tau}\right)=m,N_2\left(\hat{\tau}\right)=n,S\left(\hat{\tau}\right)=2,U\left(\hat{\tau}\right)=\hat{u},$$
$$\hat{v}<V\left(\hat{\tau}\right)\leq\hat{v}+d\,\hat{v}\},m\geq 1,n\geq 0,\hat{u}\geq 0,\hat{v}\geq 0.$$

No customer is in queue.

$$P_0\left(\hat{\tau}\right)= \Pr\{N_1\left(\hat{\tau}\right)=m,N_2\left(\hat{\tau}\right)=n,S\left(\hat{\tau}\right)=0\},m\geq 0,n\geq 0.$$

The equations governing the model are obtained as given below:

$$\frac{\partial P_0\left(\hat{\tau}\right)}{\partial\hat{t}}+\left(\lambda_1+\lambda_2\theta\right)P_0\left(\hat{\tau}\right)=\int_0^\infty Q_{0,1,1}\left(\hat{u},\hat{\tau}\right)\mu_2\left(\hat{u}\right)d\hat{u}+\int_0^\infty P_{1,0,1}\left(\hat{u},\hat{\tau}\right)\mu_1\left(\hat{u}\right)d\hat{u} \qquad (9.1)$$

$$\frac{\partial P_{m,n,1}\left(\hat{u},\hat{\tau}\right)}{\partial\hat{t}}+\frac{\partial P_{m,n,1}\left(\hat{u},\hat{\tau}\right)}{\partial\hat{u}}+\left(\lambda_1+\lambda_2\theta+\mu_1\left(\hat{u}\right)+\alpha_1\right)P_{m,n,1}\left(\hat{u},\hat{\tau}\right)$$

$$=\int_0^\infty\beta_1\left(\hat{v}\right)P_{m,n,2}\left(\hat{u},\hat{v},\hat{\tau}\right)d\hat{v}+\lambda_1\sum_{k=1}^{m-1}P_{m-k,n,1}\left(\hat{u},\hat{\tau}\right)g_1\left(m-k\right) \qquad (9.2)$$

$$+\lambda_2\theta\sum_{k=0}^{n-1}P_{m,n-k,1}\left(\hat{u},\hat{\tau}\right)g_2\left(n-k\right)$$

$$\frac{\partial P_{m,n,2}\left(\hat{u},\hat{v},\hat{\tau}\right)}{\partial\hat{t}}+\frac{\partial P_{m,n,2}\left(\hat{u},\hat{v},\hat{\tau}\right)}{\partial\hat{v}}+\left(\lambda_1+\lambda_2\theta+\beta_1\left(\hat{v}\right)\right)P_{m,n,2}\left(\hat{u},\hat{v},\hat{\tau}\right)$$

$$=\lambda_1\sum_{k=1}^{m-1}P_{m-k,n,2}\left(\hat{u},\hat{v},\hat{\tau}\right)g_1\left(m-k\right)+\lambda_2\theta\sum_{k=0}^{n-1}P_{m,n-k,2}\left(\hat{u},\hat{v},\hat{\tau}\right)g_2\left(n-k\right) \qquad (9.3)$$

$$\frac{\partial Q_{m,n,1}\left(\hat{u},\hat{\tau}\right)}{\partial \hat{t}} + \frac{\partial Q_{m,n,1}\left(\hat{u},\hat{\tau}\right)}{\partial \hat{u}} + \left(\lambda_1 + \lambda_2\theta + \mu_2\left(\hat{u}\right) + \alpha_2\right)Q_{m,n,1}\left(\hat{u},\hat{\tau}\right)$$

$$= \int_0^\infty \beta_2\left(\hat{v}\right)Q_{m,n,2}\left(\hat{u},\hat{v},\hat{\tau}\right)d\hat{v} + \lambda_1\sum_{k=1}^{m-1}Q_{m-k,n,1}\left(\hat{u},\hat{\tau}\right)g_1\left(m-k\right) \qquad (9.4)$$

$$+ \lambda_2\theta\sum_{k=0}^{n-1}Q_{m,n-k,1}\left(\hat{u},\hat{\tau}\right)g_2\left(n-k\right)$$

$$\frac{\partial Q_{m,n,2}\left(\hat{u},\hat{v},\hat{\tau}\right)}{\partial \hat{t}} + \frac{\partial Q_{m,n,2}\left(\hat{u},\hat{v},\hat{\tau}\right)}{\partial \hat{v}} + \left(\lambda_1 + \lambda_2\theta + \beta_2\left(\hat{v}\right)\right)Q_{m,n,2}\left(\hat{u},\hat{v},\hat{\tau}\right)$$

$$= \lambda_1\sum_{k=1}^{m-1}Q_{m-k,n,2}\left(\hat{u},\hat{v},\hat{\tau}\right)g_1\left(m-k\right) + \lambda_2\theta\sum_{k=0}^{n-1}Q_{m,n-k,2}\left(\hat{u},\hat{v},\hat{\tau}\right)g_2\left(n-k\right) \qquad (9.5)$$

It follows from the above definitions that $P_{m,n,1}\left(\hat{u},\hat{\tau}\right)=0$, for m < 1, n < 0, and $Q_{m,n,1}\left(\hat{u},\hat{\tau}\right)=0$, for m < 0, n < 1. We solve Eqs. (9.1)–(9.5) using the following boundary conditions:

$$P_{m,n,1}\left(0,\hat{\tau}\right) = \int_0^\infty P_{m+1,n,1}\left(\hat{u},\hat{\tau}\right)\mu_1\left(\hat{u}\right)d\hat{u} + \int_0^\infty Q_{m,n,1}\left(\hat{u},\hat{\tau}\right)\mu_2\left(\hat{u}\right)d\hat{u}$$

$$+ \delta_{0n}\lambda_1\sum_{k=0}^{m-1}P_0\left(\hat{\tau}\right)g_1\left(m-k\right) \qquad (9.6)$$

$$Q_{0,n,1}\left(0,\hat{\tau}\right) = \int_0^\infty P_{1,n,1}\left(\hat{u},\hat{\tau}\right)\mu_1\left(\hat{u}\right)d\hat{u} + \int_0^\infty Q_{0,n+1,1}\left(\hat{u},\hat{\tau}\right)\mu_2\left(\hat{u}\right)d\hat{u}$$

$$+ \lambda_2\theta\sum_{k=0}^{n-1}P_0\left(\hat{\tau}\right)g_2\left(n-k\right) \qquad (9.7)$$

$$P_{m,n,2}\left(\hat{u},0,\hat{\tau}\right) = \alpha_1 P_{m,n,1}\left(\hat{u},\hat{\tau}\right) \qquad (9.8)$$

$$Q_{m,n,2}\left(\hat{u},0,\hat{\tau}\right) = \alpha_2 Q_{m,n,1}\left(\hat{u},\hat{\tau}\right) \qquad (9.9)$$

The normalization condition is

$$P_0 + \sum_{m=1}^\infty\sum_{n=0}^\infty\left\{\int_0^\infty P_{m,n,1}\left(\hat{u}\right)d\hat{u} + \int_0^\infty\int_0^\infty P_{m,n,2}\left(\hat{u},\hat{v}\right)d\hat{u}d\hat{v}\right\}$$

$$+ \sum_{m=0}^\infty\sum_{n=1}^\infty\left\{\int_0^\infty Q_{m,n,1}\left(\hat{u}\right) + \int_0^\infty\int_0^\infty Q_{m,n,2}\left(\hat{u},\hat{v}\right)d\hat{u}d\hat{v}\right\} = 1$$

The initial condition is given by

$$P_0(0) = 1.$$

The following probability generating functions (PGFs) are used for the analysis:

$$P_1\left(\hat{z}_1,\hat{z}_2,\hat{u},\hat{\tau}\right) = \sum_{m=1}^{\infty}\sum_{n=0}^{\infty}\hat{z}_1^m\hat{z}_2^n P_{m,n,1}\left(\hat{u},\hat{\tau}\right);$$

$$Q_1\left(\hat{z}_1,\hat{z}_2,\hat{u},\hat{\tau}\right) = \sum_{m=0}^{\infty}\sum_{n=1}^{\infty}\hat{z}_1^m\hat{z}_2^n Q_{m,n,1}\left(\hat{u},\hat{\tau}\right);$$

$$P_2\left(\hat{z}_1,\hat{z}_2,\hat{u},\hat{v},\hat{\tau}\right) = \sum_{m=1}^{\infty}\sum_{n=0}^{\infty}\hat{z}_1^m\hat{z}_2^n P_{m,n,2}\left(\hat{u},\hat{v},\hat{\tau}\right);$$

$$Q_2\left(\hat{z}_1,\hat{z}_2,\hat{u},\hat{v},\hat{\tau}\right) = \sum_{m=0}^{\infty}\sum_{n=1}^{\infty}\hat{z}_1^m\hat{z}_2^n Q_{m,n,2}\left(\hat{u},\hat{v},\hat{\tau}\right).$$

We will also use Laplace transform, as defined below:

$$f_0^* = f_0^*(s) = \int_0^{\infty}\exp\left(-s\hat{\tau}\right)f_0\left(\hat{\tau}\right)d\hat{\tau}\,\mathrm{Re}(s) > 0$$

and $P_0^*(s) = \int_0^{\infty}\exp\left(-s\hat{\tau}\right)P_0\left(\hat{\tau}\right)d\hat{\tau},\ \overline{f_0^*}(s)\underline{\Delta}1 - f_0^*(s)$

The generating functions are established by solving the set of Eqs. (9.1)–(9.9) after taking Laplace transforms.

Theorem 9.1. The distributions of the server states are established in terms of the generating functions given by

$$P_1^*\left(\hat{z}_1,\hat{z}_2,\hat{u},s\right) = P_p\left(\hat{z}_1,\hat{z}_2,0,s\right)\exp\left[-\left\{Z(\lambda_1,\lambda_2)+\alpha_1\overline{r_1^*}Z(\lambda_1,\lambda_2)+s\right\}\hat{u}\right]\overline{B_1}\left(\hat{u}\right) \quad (9.10)$$

$$Q_1^*\left(\hat{z}_1,\hat{z}_2,\hat{u},s\right) = Q_q\left(\hat{z}_1,\hat{z}_2,0,s\right)\exp\left[-\left\{Z(\lambda_1,\lambda_2)+\alpha_2\overline{r_2^*}Z(\lambda_1,\lambda_2)+s\right\}\hat{u}\right]\overline{B_2}\left(\hat{u}\right) \quad (9.11)$$

$$P_2^*\left(\hat{z}_1,\hat{z}_2,\hat{u},\hat{v},s\right) = \alpha_1 P_p\left(\hat{z}_1,\hat{z}_2,0,s\right)\exp\left[-Z(\lambda_1,\lambda_2)\hat{v}\right]\overline{R_1}\left(\hat{v}\right)$$

$$\times\exp\left[-\left\{Z(\lambda_1,\lambda_2)+\alpha_1\overline{r_1^*}Z(\lambda_1,\lambda_2)+s\right\}\hat{v}\right]\overline{B_1}\left(\hat{u}\right) \quad (9.12)$$

$$Q_2^*\left(\hat{z}_1,\hat{z}_2,\hat{u},\hat{v},s\right)=\alpha_2 Q_q\left(\hat{z}_1,\hat{z}_2,0,s\right)\exp\left[-\mathbb{Z}(\lambda_1,\lambda_2)\hat{v}\right]\overline{R_2}\left(\hat{v}\right)$$

$$\times\exp\left[-\left\{\mathbb{Z}(\lambda_1,\lambda_2)+\alpha_2\overline{r_2^*}\mathbb{Z}(\lambda_1,\lambda_2)+s\right\}\hat{v}\right]\overline{B_2}\left(\hat{u}\right) \tag{9.13}$$

where

$$Q_q\left(\hat{z}_1,\hat{z}_2,0,s\right)=\frac{1-P_0^*\left(s\right)\left[\mathbb{Z}(\lambda_{z_1},\lambda_{z_2})+s\right]}{1-\left(\dfrac{1}{\hat{z}_2}\right)b_2^*\left[\mathbb{Z}(\lambda_{z_1},\lambda_{z_2})+\alpha_1\overline{r_1^*}\mathbb{Z}(\lambda_{z_1},\lambda_{z_2})+s\right]}$$

$$P_p\left(\hat{z}_1,\hat{z}_2,0,s\right)=\frac{1-P_0^*\left(s\right)\left[\mathbb{Z}(\lambda_1,\lambda_2)+s\right]}{1-\left(\dfrac{1}{\hat{z}_1}\right)b_1^*\left[\mathbb{Z}(\lambda_1,\lambda_2)+\alpha_1\overline{r_1^*}\mathbb{Z}(\lambda_1,\lambda_2)+s\right]}$$

$$-Q_q\left(\hat{z}_1,\hat{z}_2,0,s\right)\frac{1-\left(\dfrac{1}{\hat{z}_2}\right)b_2^*\left[\mathbb{Z}(\lambda_1,\lambda_2)+\alpha_2\overline{r_2^*}\mathbb{Z}(\lambda_1,\lambda_2)+s\right]}{1-\left(\dfrac{1}{\hat{z}_1}\right)b_1^*\left[\mathbb{Z}(\lambda_1,\lambda_2)+\alpha_1\overline{r_1^*}\mathbb{Z}(\lambda_1,\lambda_2)+s\right]}$$

$$P_0^*\left(s\right)=\left[\lambda_1\left(1-\overline{G_1}\left(\hat{z}_1\left(\hat{z}_{2s},s\right)\right)\right)+\lambda_2\theta\left(1-\overline{G_2}\left(\hat{z}_{2s}\right)\right)\right.$$

$$\left.+\alpha_1\overline{r_1^*}\left(\lambda_1\left(1-\overline{G_1}\left(\hat{z}_1\left(\hat{z}_{2s},s\right)\right)\right)+\lambda_2\theta\left(1-\overline{G_2}\left(\hat{z}_{2s}\right)\right)\right)+s\right]^{-1}$$

$$\mathbb{Z}(\lambda_1,\lambda_2)=\left(\lambda_1\left(1-\overline{G_1}\left(\hat{z}_1\right)\right)+\lambda_2\theta\left(1-\overline{G_2}\left(\hat{z}_2\right)\right)\right)$$

$$\mathbb{Z}(\lambda_{z_1},\lambda_{z_2})=\lambda_1\left(1-\overline{G_1}\left(\hat{z}_1\left(\hat{z}_2,s\right)\right)\right)+\lambda_2\theta\left(1-\overline{G_2}\left(\hat{z}_2\right)\right)$$

Lemma 9.1

    i. *When the server busy.*

      The marginal generating functions (MGFs) of the category 1 and category 2 customers are given by

$$P_1^*\left(\hat{\mathbf{z}}_1,\hat{\mathbf{z}}_2,s\right)=P_p\left(\hat{\mathbf{z}}_1,\hat{\mathbf{z}}_2,0,s\right)\frac{1-b_1^*\left[\mathbb{Z}(\lambda_1,\lambda_2)+\alpha_1\overline{r_1^*}\ \mathbb{Z}(\lambda_1,\lambda_2)+s\right]}{\left[\mathbb{Z}(\lambda_1,\lambda_2)+\alpha_1\overline{r_1^*}\ \mathbb{Z}(\lambda_1,\lambda_2)+s\right]} \tag{9.14}$$

$$Q_1^*\left(\hat{\mathbf{z}}_1,\hat{\mathbf{z}}_2,s\right)=Q_q\left(\hat{\mathbf{z}}_1,\hat{\mathbf{z}}_2,0,s\right)\frac{1-b_2^*\left[\mathbb{Z}(\lambda_1,\lambda_2)+\alpha_2\overline{r_2^*}\ \mathbb{Z}(\lambda_1,\lambda_2)+s\right]}{\left[\mathbb{Z}(\lambda_1,\lambda_2)+\alpha_2\overline{r_2^*}\ \mathbb{Z}(\lambda_1,\lambda_2)+s\right]} \tag{9.15}$$

ii. **When the server down.**

In this case, marginal generating functions (MGFs) of the category 1 and category 2 customers are given by

$$P_2^*\left(\hat{z}_1,\hat{z}_2,s\right)=\alpha_1 P_p\left(\hat{z}_1,\hat{z}_2,0,s\right)\frac{1-b_1^*\left[Z(\lambda_1,\lambda_2)+\alpha_1\overline{r_1^*}Z(\lambda_1,\lambda_2)+s\right]}{\left[Z(\lambda_1,\lambda_2)+\alpha_1\overline{r_1^*}Z(\lambda_1,\lambda_2)+s\right]}$$

$$\times\frac{1-r_1^*\left[Z(\lambda_1,\lambda_2)+s\right]}{\left[Z(\lambda_1,\lambda_2)+s\right]} \tag{9.16}$$

$$Q_2^*\left(\hat{z}_1,\hat{z}_2,s\right)=\alpha_2 Q_q\left(\hat{z}_1,\hat{z}_2,0,s\right)\frac{1-b_2^*\left[Z(\lambda_1,\lambda_2)+\alpha_2\overline{r_2^*}\,Z(\lambda_1,\lambda_2)+s\right]}{\left[Z(\lambda_1,\lambda_2)+\alpha_2\overline{r_2^*}\,Z(\lambda_1,\lambda_2)+s\right]}$$

$$\times\frac{1-r_2^*\left[Z(\lambda_1,\lambda_2)+s\right]}{\left[Z(\lambda_1,\lambda_2)+s\right]} \tag{9.17}$$

Lemma 9.2

The joint PGF (probability generating function) for the number of both categories customers in the system is

$$P^*\left(\hat{z}_1,\hat{z}_2,s\right)=P_1^*\left(\hat{z}_1,\hat{z}_2,s\right)+P_2^*\left(\hat{z}_1,\hat{z}_2,s\right)+Q_1^*\left(\hat{z}_1,\hat{z}_2,s\right)+Q_2^*\left(\hat{z}_1,\hat{z}_2,s\right)$$

$$=P_p\left(\hat{z}_1,\hat{z}_2,0,s\right)\frac{\overline{b_1^*}\left[Z(\lambda_1,\lambda_2)+\alpha_1\overline{r_1^*}\,Z(\lambda_1,\lambda_2)+s\right]}{\left[Z(\lambda_1,\lambda_2)+s\right]}$$

$$+Q_q\left(\hat{z}_1,\hat{z}_2,0,s\right)\frac{\overline{b_2^*}\left[Z(\lambda_1,\lambda_2)+\alpha_2\overline{r_2^*}\,Z(\lambda_1,\lambda_2)+s\right]}{\left[Z(\lambda_1,\lambda_2)+s\right]}. \tag{9.18}$$

Lemma 9.3

The limiting behavior of the joint distribution is given by the joint probability generating function

$$P\left(\hat{z}_1,\hat{z}_2\right)=\frac{P_0\,\Delta_1\left(\hat{z}_1,\hat{z}_2\right)\left(1-1/\hat{z}_1\right)}{1-\left(1/\hat{z}_1\right)\Delta_1\left(\hat{z}_1,\hat{z}_2\right)}+Q_q\left(\hat{z}_1,\hat{z}_2,0\right)$$

$$\frac{1-\Delta_2\left(\hat{z}_1,\hat{z}_2\right)-\left[1-\Delta_2\left(\hat{z}_1,\hat{z}_2\right)\right]\left[1-\Delta_1\left(\hat{z}_1,\hat{z}_2\right)\right]/\left[1-\Delta_1\left(\hat{z}_1,\hat{z}_2\right)/\hat{z}_1\right]}{\left[\mathbb{Z}(\lambda_1,\lambda_2)\right]} \tag{9.19}$$

where

$$\Delta_1\left(\hat{z}_1,\hat{z}_2\right)=b_1^*\left[\mathbb{Z}(\lambda_1,\lambda_2)+\alpha_1\overline{r_1^*}\mathbb{Z}(\lambda_1,\lambda_2)\right],\Delta_2\left(\hat{z}_1,\hat{z}_2\right)$$

$$=b_2^*\left[\mathbb{Z}(\lambda_1,\lambda_2)+\alpha_2\overline{r_2^*}\mathbb{Z}(\lambda_1,\lambda_2)\right]$$

$$Q_q\left(\hat{z}_1,\hat{z}_2,0\right)=-\frac{P_0\ \mathbb{Z}(\lambda_1,\lambda_2)}{1-\left(1/\hat{z}_2\right)b_2^*\left[\mathbb{Z}(\lambda_1,\lambda_2)+\alpha_2\overline{r_2^*}\ \mathbb{Z}(\lambda_1,\lambda_2)\right]}$$

and $z_1(z_2)$ is the smallest root of the equation

$$\hat{z}_1=b_1^*\left[\mathbb{Z}(\lambda_1,\lambda_2)+\alpha_1\overline{r_1^*}\ \mathbb{Z}(\lambda_1,\lambda_2)\right].$$

9.4   PERFORMANCE INDICIES

For the quantitative assessment of performance of concern queueing system, we shall derive explicit expression for mean queue lengths and average waiting time of priority and nonpriority customers as follows:

i. **Average queue length**
   Theorem 9.2. The mean queue length of the priority (L_p) and non-priority customers (L_q) are

$$L_p=\rho_1\left(1+\alpha_1\gamma_1'\right)$$

$$+\frac{\lambda_1\widehat{g_1}\left[\lambda_1\widehat{g_1}\left(1+\alpha_1\gamma_1'\right)^2\tau_1''+\lambda_2\widehat{g_2}\left(1+\alpha_2\gamma_2'\right)^2\tau_2''\right]+\left[\widehat{g_1''}\rho_1\left(1+\alpha_1\gamma_1'\right)/\widehat{g_1}\right]+\left[\lambda_1\widehat{g_1}\left(\alpha_1\rho_1\gamma'+\alpha_2\rho_2\gamma_2'\right)\right]}{2\left(1-\rho_1\left(1+\alpha_1\gamma_1'\right)\right)}$$

(9.20)

$$L_q=\rho_2\left(1+\alpha_2\gamma_2'\right)$$

$$+\frac{\lambda_2\widehat{g_2}\left[\lambda_1\widehat{g_1}\left(1+\alpha_1\gamma_1'\right)^2\tau_1''+\lambda_2\widehat{g_2}\left(1+\alpha_2\gamma_2'\right)^2\tau_2''+\left(\widehat{g_1''}\rho_1\tau_1'/\widehat{g_1}\right)\left(1+\alpha_1\gamma_1'\right)^2\right]}{2\left[1-\rho_1\left(1+\alpha_1\gamma_1'\right)\right]\left[1-\rho_1\left(1+\alpha_1\gamma_1'\right)-\rho_2\left(1+\alpha_2\gamma_2'\right)\right]}$$

$$+\frac{\left[\left(\widehat{g_2''}\rho_2/\widehat{g_2}\right)\left(1+\alpha_1\gamma_1'\right)\right]\left[1-\rho_1\left(1+\alpha_1\gamma_1'\right)\right]+\alpha_2\lambda_2\rho_2\gamma_2''}{2\left[1-\rho_1\left(1+\alpha_1\gamma_1'\right)\right]\left[1-\rho_1\left(1+\alpha_1\gamma_1'\right)-\rho_2\left(1+\alpha_2\gamma_2'\right)\right]}$$

$$\frac{-\alpha_2\lambda_2\widehat{g_2}\gamma_2''\rho_1^2\rho_2\left(1+\alpha_1\gamma_1'\right)^2-\alpha_2\lambda_2\widehat{g'}_2\ _2\gamma_2''\rho_2^2\rho_1\left(1+\alpha_1\gamma_1'\right)\left(1+\alpha_2\gamma_2'\right)-\alpha_1\lambda_2\widehat{g'}_2\rho_1\gamma_1'}{}$$

(9.21)

ii. Mean waiting time

Little's formula is employed to obtain the mean waiting times (MWTs) of priority and nonpriority customers in the steady state. Thus

$$W_p = \frac{L_p}{\lambda_1 \widehat{g_1'}} = \tau_1' \left(1 + \alpha_1 \gamma_1'\right)$$

$$+ \frac{\tau_1' \left[\lambda_1 \widehat{g_1'} \left(1 + \alpha_1 \gamma_1'\right)^2 \tau_1'' + \lambda_2 \widehat{g_2'} \left(1 + \alpha_2 \gamma_2'\right)^2 \tau_2''\right] + \left[\widehat{g_1''} \tau'\left(1 + \alpha_1 \gamma_1'\right) / \widehat{g_1'}\right]}{2\left(1 - \rho_1 \left(1 + \alpha_1 \gamma_1'\right)\right)} \quad (9.22)$$

$$\frac{+ \left[\tau_1' \left(\alpha_1 \rho_1 \gamma' + \alpha_2 \rho_2 \gamma_2'\right)\right]}{}$$

$$W_q = \frac{L_q}{\lambda_2 \widehat{g_2'}} = \tau_2' \left(1 + \alpha_2 \gamma_2'\right)$$

$$+ \frac{\splitfrac{\tau_2'\left[\lambda_1 \widehat{g_1'}\left(1+\alpha_1\gamma_1'\right)^2 \tau_1'' + \lambda_2 \widehat{g_{22}'}\left(1+\alpha_2\gamma_2'\right)^2 \tau_2'' + \left(\widehat{g_1''}\rho_1\tau_1' / \widehat{g_1'}\right)\left(1+\alpha_1\gamma_1'\right)^2\right]}{+\left[\left(\widehat{g_2''}\tau_2' / \widehat{g_2'}\right)\left(1+\alpha_1\gamma_1'\right)\right]\left[1-\rho_1\left(1+\alpha_1\gamma_1'\right)\right] + \alpha_2\lambda_2\tau_2'\gamma_2'' - \alpha_2\gamma_2'' \rho_1^2\rho_2\left(1+\alpha_1\gamma_1'\right)^2}}{\splitfrac{- \alpha_2\gamma_2'' \rho_2^2\rho_1\left(1+\alpha_1\gamma_1'\right)\left(1+\alpha_2\gamma_2'\right) - \alpha_1\rho_1\gamma_1'}{2\left[1-\rho_1\left(1+\alpha_1\gamma_1'\right)\right]\left[1-\rho_1\left(1+\alpha_1\gamma_1'\right)-\rho_2\left(1+\alpha_2\gamma_2'\right)\right]}}}$$

$$(9.23)$$

9.5 SPECIAL CASES

Now we discuss few existing results appeared in the literature which are special cases of our model.

a. When $\alpha_1 = \alpha_2 = 0$, and $\theta = 1$, our model provides results for $M_1^{X_1} M_2^{X_2} / G_1 G_2 / 1$, model discussed in Chaudhry and Templeton (1983). In this case Eqs. (9.10) and (9.11) reduce to

$$P_1^* \left(\hat{z}_1, \hat{z}_2, \hat{u}, s\right) = P_p \left(\hat{z}_1, \hat{z}_2, 0, s\right) \exp\left[-\mathbb{Z}(\lambda_1, \lambda_2) + s\right] \quad (9.24)$$

$$Q_1^* \left(\hat{z}_1, \hat{z}_2, \hat{u}, s\right) = Q_q \left(\hat{z}_1, \hat{z}_2, 0, s\right) \exp\left[-\mathbb{Z}(\lambda_1, \lambda_2) + s\right] \quad (9.25)$$

Now, the expectations of queue size of priority and nonpriority customers respectively, are

$$L_p = \rho_1 + \frac{\lambda_1 \widehat{g_1'} \left[\lambda_1 \widehat{g_1'}\tau_1'' + \lambda_2 \widehat{g'}_2 \tau_2''\right] + \left[\widehat{g_1''} \rho_1 / \widehat{g_1'}\right]}{2\left(1 - \rho_1\right)} \quad (9.26)$$

$$L_q = \rho_2 + \frac{\lambda_2 \widehat{g_2'}\left[\lambda_1 \widehat{g_1'}\tau_1'' + \lambda_2 \widehat{g_2'}\tau_2'' + \left(\widehat{g_1''}\rho_1\tau_1' / \widehat{g_1'}\right)\right] + \left(\widehat{g_2''}\rho_2 / \widehat{g_2'}\right)\left(1-\rho_1\right)}{2\left(1-\rho_1\right)\left(1-\rho_1-\rho_2\right)} \quad (9.27)$$

b. When $\alpha_1 = \alpha_2 = 0$, $\lambda_2 = 0$, $\hat{z}_2 = 0$, $b_2(\hat{u}) = 0$, there are no nonpriority customers in the queue. In this case, Eq. (9.14) reduces to

$$P(\hat{z}_1) = \frac{P_0\, b_1^*\left(\lambda_1 - \lambda_1 \bar{G}_1(\hat{z}_1)\right)(\hat{z}_1 - 1)}{\hat{z}_1 - b_1^*\left(\lambda_1 - \lambda_1 \bar{G}_1(\hat{z}_1)\right)}, \quad \text{where } P_0 = 1 - \rho_1 \qquad (9.28)$$

Now, we obtain

$$L_p = \rho_1 + \frac{\left(\lambda_1 \widehat{g_1'}\right)^2 \tau_1'' + \left(\widehat{g_1''}\rho_1 / \widehat{g_1'}\right)}{2(1 - \rho_1)} \qquad (9.29)$$

c. When $\alpha_1 = \alpha_2 = 0$, $\lambda_1 = 0$, $\theta = 1$, $\hat{z}_1 = 0$, $b_1(\hat{u}) = 0$, there are no priority customers in the queue, so that Eq. (9.16) reduce to

$$P(\hat{z}_2) = \frac{P_0\left[1 - b_2^*\left(\lambda_2 - \lambda_2 \bar{G}_2(\hat{z}_2)\right)\right]\hat{z}_2}{\left[\hat{z}_2 - b_2^*\left(\lambda_2 - \lambda_2 \bar{G}_2(\hat{z}_2)\right)\right]}, \quad \text{where } P_0 = 1 - \rho_2 \qquad (9.30)$$

Thus, we obtain

$$L_q = \rho_2 + \frac{\left(\lambda_2 \widehat{g_2'}\right)^2 \tau_2'' + \left(\widehat{g_2''}\rho_2 / \widehat{g_2'}\right)}{2(1 - \rho_2)} \qquad (9.31)$$

9.6   SENSITIVITY ANALYSIS

All-encompassing numerical experimentation has been done to study the effects of failure and repair rates on the expected queue length for priority (L_p) and nonpriority (L_q) customers by varying different parameters as shown in Figures 9.1–9.7. The

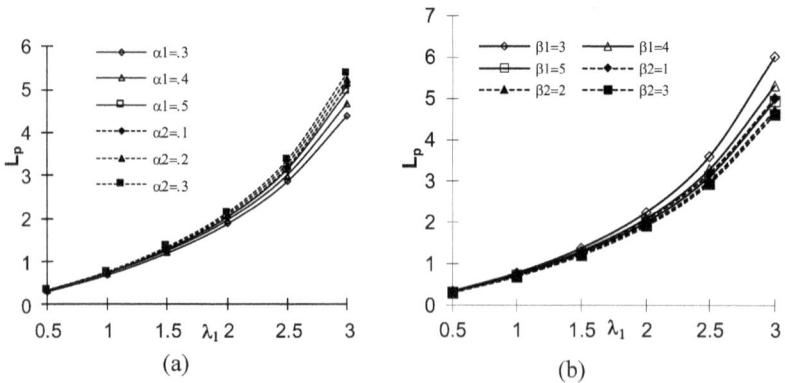

FIGURE 9.1   Expected queue lengths for priority (L_p) customers vs. λ_1 for varying parameters of (a) (α_1 and α_2) (b) (β_1 and β_2).

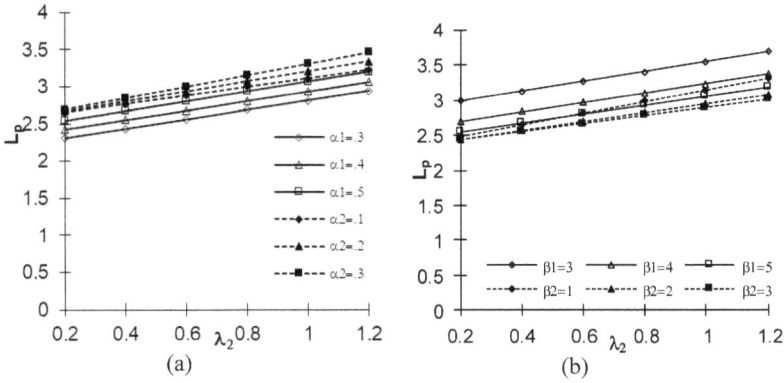

FIGURE 9.2 Expected queue lengths for priority (L_p) customers vs. λ_2 for varying parameters of (a) (α_1 and α_2) (b) (β_1 and β_2).

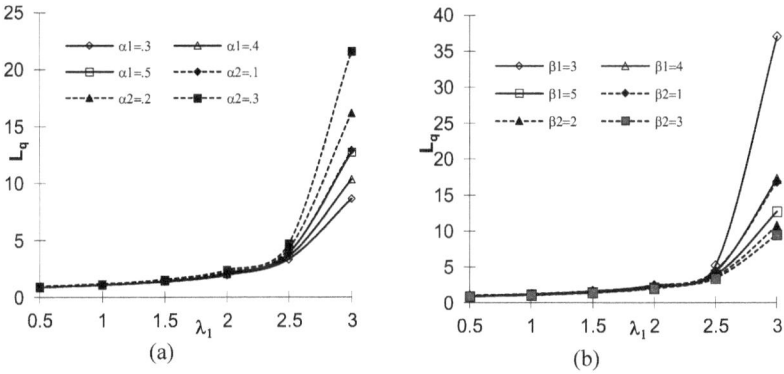

FIGURE 9.3 Expected queue lengths for nonpriority (L_q) customers vs. λ_1 for varying parameters of failure rate and repair rate (a) (α_1 and α_2) (b) (β_1 and β_2).

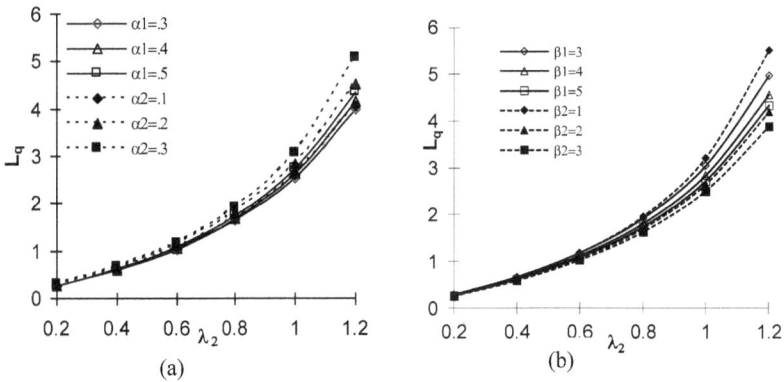

FIGURE 9.4 Expected queue lengths for nonpriority (L_q) customers vs. λ_2 for varying parameters of failure rate and repair rate (a) (α_1 and α_2) (b) (β_1 and β_2).

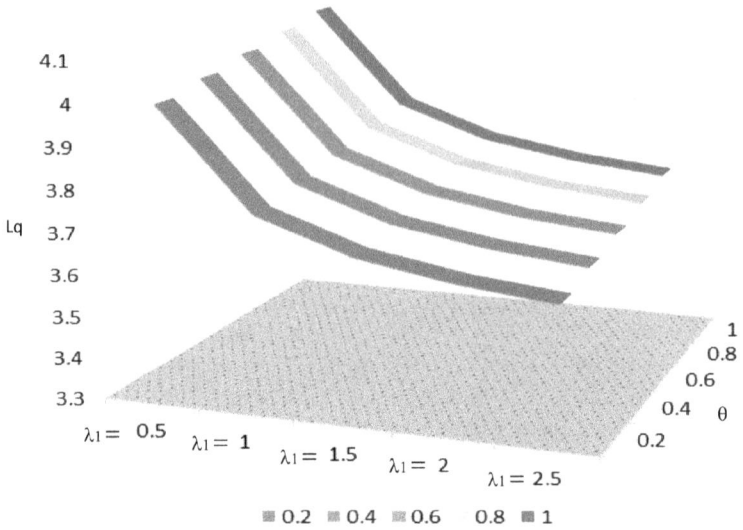

FIGURE 9.5 Expected queue lengths for nonpriority (L_q) customers vs. λ_1 for varying balking parameter θ.

numerical program has been developed in computer software MATLAB and run on Window 8 to implement the computational procedure.

Figures 9.1–9.4 depict the impact of arrival rates λ_1 and λ_2 on the expected queue lengths for priority and nonpriority customers for different values of failure rates (α_1, α_2) and the repair rates (β_1, β_2). For numerical computation, we fix other parameters as $\mu_1 = 5$, $\mu_2 = 2$, $\alpha_1 = 0.5$, $\beta_1 = 5$, $\alpha_2 = 0.2$, $\beta_2 = 3$, $\theta = 0.2$. The expected queue lengths for priority and nonpriority customers sharply increase with the arrival rates

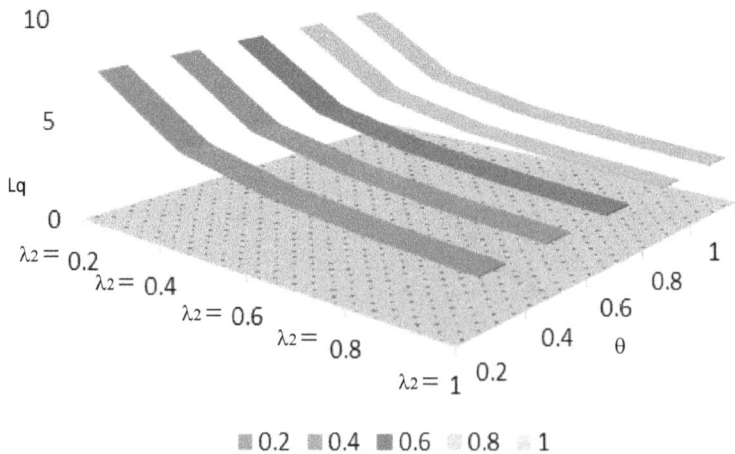

FIGURE 9.6 Expected queue lengths for nonpriority (L_q) customers vs. λ_2 for varying balking parameter θ.

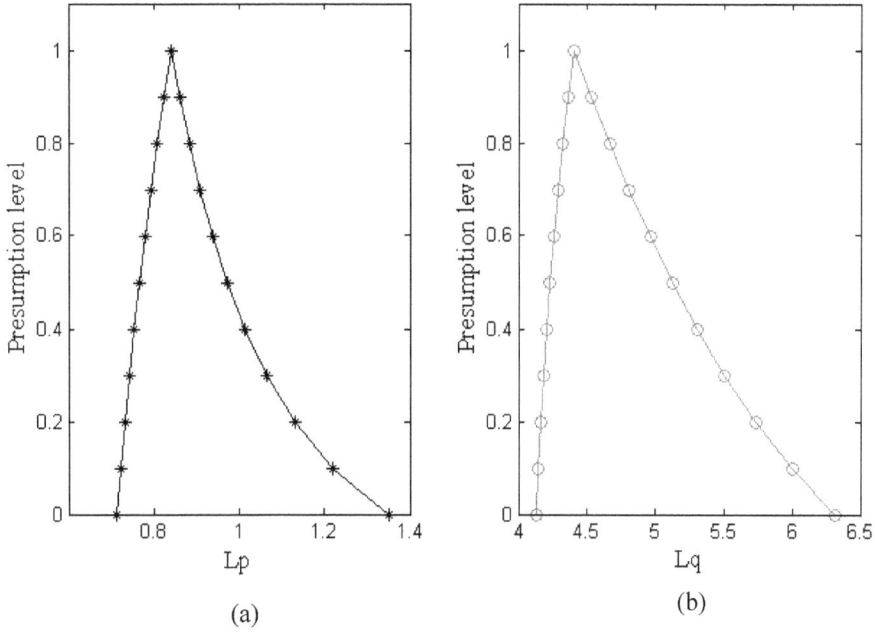

FIGURE 9.7 Fuzzy membership function of (a) priority customers (Lp) (b) nonpriority customers (Lq).

λ_1 and λ_2, but a linear increment for L_p is observed with increasing values of λ_2. We also examine the effect of failure rates (α_1, α_2) and the repair rates (β_1, β_2) and observe that the expected queue length increases as failure rates (α_1, α_2) increase but decreases as repair rates (β_1, β_2) increase for both categories of customers.

Figures 9.5 and 9.6 exhibit the trends of expected queue sizes for priority and non-priority customers for different values of λ_1 and λ_2 by varying parameters θ. The increments in the values of λ_1 and λ_2 result in significant decrement in the expected queue size for both priority and nonpriority types of customers. The expected queue sizes for priority and nonpriority customers increase gradually as balking parameter increases.

9.6.1 Fuzzy Results

The proposed model has been tested by considering the crisp and fuzzy values of parameters. The sensitivity analysis of the proposed model in case of crisp parameters is taken into account. In this section, uncertainty involved in parameters has been modeled using triangular fuzzy numbers (Pachauri et al., 2013). The fuzzy values of the parameters are taken as follows:

$$\mu_1 = [4.75; 5; 5.25], \mu_2 = [1.9; 2; 2.1], \alpha_1 = [0.475; 0.5; 0.525], \beta_1 = [4.75; 5; 5.25],$$

$$\alpha_2 = [0.19; 0.2; 0.21], \beta_2 = [2.85; 3; 3.15], \lambda_1 = [0.5; 1.75; 3],$$

$$\lambda_2 = [1; 1.1; 1.2] \text{ and } \theta = 0.2.$$

TABLE 9.1

Upper and Lower Limits of L_p

Presumption Level	Upper Limit	Lower Limit
0.0	1.3508	0.7119
0.1	1.2197	0.7217
0.2	1.1301	0.7319
0.3	1.0642	0.7427
0.4	1.0132	0.7541
0.5	0.9721	0.7662
0.6	0.9380	0.7791
0.7	0.9091	0.7930
0.8	0.8840	0.8080
0.9	0.8618	0.8242
1.0	0.8421	0.8421

By using the arithmetic operations on fuzzy numbers, expected queue lengths for priority (L_p) and nonpriority (L_q) customers have been calculated. The upper and lower limits of L_p and L_q are given in Tables 9.1 and 9.2. Fuzzy membership functions of L_p and L_q are defined in Figure 9.7. From Tables 9.1 and 9.2, it is noted that expected queue length for priority (L_p) customers varies between 0.7 and 1.3. Since decimal value of L_p is not possible, we will use flour function. Accordingly, the minimum value of L_p is 1 and maximum is 2. whereas, when crisp value of parameters is considered, L_p varies between 1 and 6.

Based on numerical computation, finally we can say that the expected queue sizes for priority (L_p) and non-priority (L_q) customers increase with the arrival rate. But on decreasing (increasing) the failure rate (repair rate), the expected queue sizes for

TABLE 9.2

Upper and Lower Limits of L_q

Presumption Level	Upper Limit	Lower Limit
0.0	6.3163	4.1323
0.1	5.9937	4.1452
0.2	5.7312	4.1610
0.3	5.5063	4.1798
0.4	5.3072	4.2016
0.5	5.1274	4.2268
0.6	4.9627	4.2554
0.7	4.8104	4.2877
0.8	4.6684	4.3240
0.9	4.5353	4.3647
1.0	4.4101	4.4101

both categories of customer's increase (decrease); the same trends we expect in real life situations. Similar interpretation can be made for the expected queue length for nonpriority (L_q) customers. The fuzzy results show that model performs better when fuzzy values of parameters are considered instead of crisp value.

9.7   CONCLUDING REMARKS

There are many congestion situations in which arrivals occur in bulk but have different priority. The ships arriving at a port in convoy, letters arriving at a post office, people going to watch movies in theatre, a family restaurant and so on, are some of the examples of queueing situations which well fitted to proposed model. The customer's behavior is very important and plays the key role in real world congestion problems. To demonstrate the use of the model, different parametric values have been specified and lucidly illustrated with the help of graphs. The numerical results provide significant insights to the system designers and decision makers.

REFERENCES

Al-Seedy, R. O., El-Sherbiny, A. A., El-Shehawy, S. A. and Ammar, S. I. 2009. Transient solution of the M/M/c queue with balking and reneging, *Computational Mathematics and Applications*. 57: 1280–1285.

Ayyappan, G., Somasundaram, B. 2019. Analysis of two stage M[X1];M[X2]/G1;G2/1 retrial G-queue with discretionary priority services, working breakdown, Bernoulli vacation, preferred and impatient units, *Applications & Applied Mathematics*. 14(2): 640–671.

Bellman, L. A. and Zadeh, R. E. 1970. Decision-making in a fuzzy environment, *Management Science*. 17: B141–B164.

Buckley, J. J. 1990. Elementary queueing theory based on possibility theory, *Fuzzy Sets and Systems*. 37: 43–52.

Buckley, J. J., Feuring, T. and Hayashi, Y. 2001. Fuzzy queueing theory revisited, *International Journal of Uncertainty, Fuzziness and Knowledge-Based Systems*. 9(5): 527–537.

Choudhury, G., Tadj, L. and Deka, K. 2010. A batch arrival retrial queueing system with two phases of service and service interruption, *Computational Mathematics and Applications*. 59(1): 437–450.

Chaudhry, M. L. and Templeton, G. O. 1983. A First Course in Bulk Queue, John Wiley and Sons, New York, NY.

Dimitriou, I. 2013a. A batch arrival priority queue with recurrent repeated demands, admission control and hybrid failure recovery discipline, *Applied Mathematical Modeling*. 219(24): 11327–11340.

Dimitriou, I. 2013b. A mixed priority retrial queue with negative arrivals, unreliable server and multiple vacations, *Applied Mathematical Modeling*. 37(3): 1295–1309.

Drekic, S. and Woolford, D. G. 2005. A preemptive priority queue with balking, *European Journal of Operational Research*. 164(2): 387–401.

Dudin, A., Kim, C., Dudin, S. and Dudina, G. 2015. Priority retrial queueing model operating in random environment with varying number and reservation of servers, *Applied Mathematics and Computations*. 269: 674–690.

Hassan, N. A. and Hoda Ibrahim, S. A. 2013. Analysis of multi-level queueing systems with servers breakdown by using recursive solution technique, *Applied Mathematical Modeling*. 37(6): 3714–3723.

Jain, A. and Ahuja, A. 2020. Study of FM/FM(FM)/1/L queue with server startup under threshold N policy using parametric non-linear programming method, Asset Analytics – Performance and Safety Management. 50–79, DOI: 10.1007/978-981-15-3643-4_5.

Jain, A. and Jain, M. 2019. Balking strategies for a working vacation priority queueing system with two classes of customers, Springer's Computer Science Proceeding in International Conference on Recent Trends in Operations Research and Statistics, 165–176, https://doi.org/10.1007/978-981-13-0857-4_12.

Jain, M. and Bhagat, A. 2013. Transient analysis of retrial queues with double orbits and priority customers. In: W.-L. Chen and C.-C. Kuo (Eds). Proc. 8th International Conference on Queuing Theory and Network Applications. Taichung, Taiwan, 235–240.

Jain, M. and Jain, A. 2014. Batch arrival priority queueing model with second optional service and server breakdown, *International Journal of Operations Research*. 11(4): 112–130.

Jain, M. and Sanga, S. S. 2020. Fuzzy cost optimization and admission control for machine interference problem with general retrial, *Journal of Testing and Evaluation*. 48, online, 01 November 2020, https://doi.org/10.1520/JTE20180882.

Kao, C., Li., C. and Chen, S. 1999. Parametric programming to the analysis of fuzzy queues, *Fuzzy Sets and Systems*. 107: 93–100.

Ke, J. B., Lee, W. C. and Wang, K. H. 2007. Reliability and sensitivity analysis of a system with multiple unreliable service stations and standby switching failures, *Physica A. Statistical Mechanics and its Applications*. 380: 455–469.

Ke, J. C. and Lin, C. H. 2006. Fuzzy analysis of queuing systems with an unreliable server. a nonlinear programming approach, *Applied Mathematics and Computation*. 175: 330–346.

Li, R. J. and Lee, E. S. 1989. Analysis of fuzzy queues, *Computers and Mathematics with Applications*. 17(7): 1143–1147.

Pachauri, B., Kumar, A. and Dhar, J. 2013. Modeling optimal release policy under fuzzy paradigm in imperfect debugging environment, *Inf. Softw. Technol.* 55(11): 1974–1980.

Prade, H. M. 1980. An outline of fuzzy or possibilistic models for queuing systems. In: P.P. Wang, S.K. Chang (Eds). Fuzzy Sets: Theory and Applications to Policy Analysis and Information Systems. Plenum Press, New York, NY, 147–153.

Sanga, S. S. and Jain, M. 2019. FM/FM/1 double orbit retrial queue with customers' joining strategy: A parametric nonlinear programing approach, *Applied Mathematics and Computation*. 362:124542 https://doi.org/10.1016/j.amc.2019.06.056.

Shekhar, C., Jain, M. and Bhatia, S. 2014. Fuzzy analysis of machine repair problem with switching failure and reboot, *Journal of Reliability and Statistical Studies*. 7: 41–55.

Singh, C. J., Kaur, S. and Jain, M. 2020. Unreliable server retrial G-queue with bulk arrival, optional additional service and delayed repair, *International Journal of Operational Research*. 38(1): 82–111.

Vadivu, A. S., Vanayak, R., Dharmaraja, S. and Arumuganathan, R. 2014. Performance analysis of voice over internet protocol via non-Markovian loss system with preemptive priority and server breakdown, *OPSERACH*. 5(1): 50–75.

10 Application of Fuzzy AHP Approach for Evaluation of Sustainable Energy Sources in India

S. K. Saraswat, Abhijeet K Digalwar and S. S. Yadav
Birla Institute of Technology & Science, Pilani, Rajasthan, India

CONTENTS

10.1 INTRODUCTION

Sustainable energy is defined as "A safe, environmentally sound, and economically viable energy pathway that will sustain human progress into the distant future is clearly imperative (Klein and Whalley 2015; Mainali et al. 2014). Energy is an essential factor for the economic development of the country and an all-important factor in human life. Per capita energy consumption will show the economic prosperity of the nation (Kahraman and Kaya 2010; Lee and Chang 2018). In during the 1970s, energy sustainability was analyzed with only a single criterion, especially economic or low cost was known as a "single-pillar" analysis (Klein and Whalley 2015; Maxim 2014). However, in the 1980s, due to technical development, a technical factor was added as a secondary criterion. In the 1990s, a growing awareness about the environment and social issues lead to modifications in the above decision framework. The necessity of accompanying the technical, environmental and social factors in sustainability analysis gave way to implementing a multi-criteria decision making (MCMD) approach (Büyüközkan and Güleryüz 2016; Kaya and Kahraman 2010).

145

The MCDM approach is used for the selection of the best single option among a set of options by evaluating them with multiple decision criteria (Arce et al. 2015; Çolak and Kaya 2017). To prioritize or rank alternatives, first select the MCDM approach, suitable criteria, subcriteria and alternatives related to their goal. Second, collect the weights of experts about the importance of criteria, subcriteria and alternatives (Amer and Daim 2011). Commonly used MCDM approaches are AHP (Analytical Hierarchy Process), WSM (Weighted Sum Method), PROMETHEE (Preference Ranking Organization Method for Enrichment Evaluations), SMART (Simple Multi-Attribute Rating Technique), ELECTRE (Elimination Choice Translating Reality), TOPSIS (Technique for Order Preference by Similarity to Ideal Solution) and ANP (Analytic Network Process) etc. Pohekar and Ramachandran (2004) reviewed more than 90 research articles and concluded that AHP, PROMEETHE and ELECTRE MCDM approaches are frequently used MCDM techniques in the sustainable energy planning sector. Amer and Daim (2011) found AHP is a widely used MCDM approach because of its flexibility and robustness. Kahraman, Kaya and Cebi (2009) wrote that AHP is the most outstanding MCDM tool to solve energy management problems. According to Çolak and Kaya (2017) fuzzy AHP is the most commonly used the approach in an energy field with a 40 percent share in overall fuzzy MCMD studies.

India is the second largest country in overall population, and it is the first largest country in the rural population in all over the world. IMF's World Economic Outlook projected India's growth rate will accelerate to 7.5 percent until 2026–2027 (Pappas and Chalvatzis 2017). Real GDP of the country raises with an increase in energy consumption, so energy acts as an essential factor for the economic development of every sector of the country (Kumar Shukla and Sharma 2017). At present, India is the third largest country in the world for both electricity generation and consumption, after China and the United States. Energy generation in India has been reached 2.5 times higher than the year 1997–98. India generates 344,689 MW of power, out of which thermal energy generates a major portion with 64.3 percent of overall generation. In the remaining portion, renewable energy covers 20.5 percent, followed by hydro energy (13.2 percent) and nuclear energy (2 percent) (Ministry of Power 2019). Energy demand is continuously increasing in India due to the development of industrialization and urbanization. Energy demand in India reached 915,123 million units in the years 2017–2018, which is double the energy requirement in 1997–1998. Figure 10.1 shows the increase in energy demand from 1997–1998 to 2017–2018. According to the government of India (Energy Statistics 2017), energy demand will rise to 551.8 trillion units until 2047. To fulfill current or future energy demand, India is mainly dependent on fossil fuels, which are a major source for greenhouse gas emission and climate change. Hydro energy is considered as clean and an ideal source for meeting India's peak power demand, but due to social acceptance of large hydroelectric dams made their use limited (Hairat and Ghosh 2017). Nuclear energy is economic and clean, but the cost of import of nuclear fuel and public agitations against nuclear power will weak the future nuclear aspect. Renewable energy sources are the clean and economic source, but they have some drawbacks, such as their reliability and intermittent nature. So, there is a necessity to evaluate the most sustainable energy source in India.

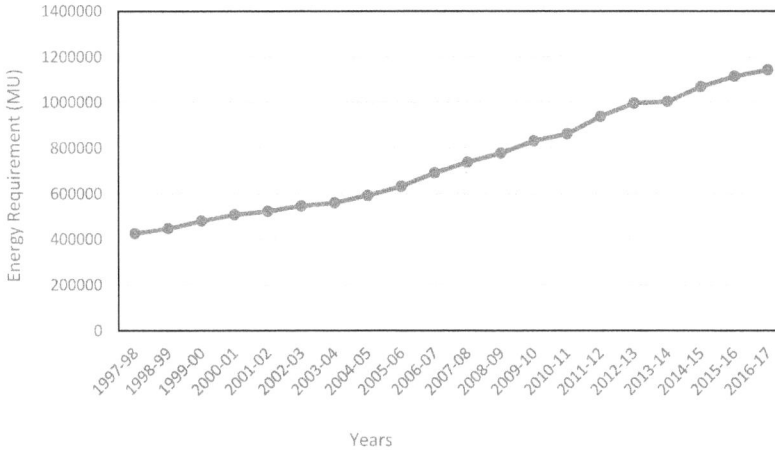

FIGURE 10.1   Trend of increase in energy requirement.

In this research work, fuzzy AHP approach is employed for evaluation of most sustainable energy source among renewable (solar, wind, biomass, hydro) and non-renewable (thermal, gas power, nuclear) energy sources. This analysis includes six criteria: environmental, technical, economic, social, political and flexibility. In the best knowledge of the author, there is no such literature available in which seven energy sources as an alternative and six criteria are considered for the analysis.

The rest of chapter is organized as follows: In section 10.2, a collected recent literature review. In section 10.3, a methodology of fuzzy AHP is explained. In section 10.4, fuzzy AHP approach is employed for the selection of the most sustainable energy source and calculation and results of the analysis are presented. In section 10.5, the conclusions of the research work are presented.

10.2   LITERATURE SURVEY

The MCMD approach is an attractive tool in decision making since last two-three decades especially in energy planning and management problems. In last few years, the MCDM (AHP) approach has been widely used in sustainable energy planning, as presented in Table 10.1. The AHP approach is also widely used with the geographical information system (GIS) tool for selection of the most suitable location for the different energy sources. Uyan (2017) combined AHP with GIS software to select the most suitable locations for solar power installation in Turkey. He analyzed these locations by covering the economic as well as environmental factors. Baseer et al. (2017) analyzed for the Kingdom of Saudi Arabia to install wind power plants. By combining AHP with GIS, they highlighted near Ras Tanura on the coast in the eastern province, Al-Wajh on the coast in the western region, Turaif in Al-Jawf region at northern borders as a most suitable sites in Saudi Arabia. Aly, Jensen and Pedersen (2017) investigated spatial suitability for large-scale solar power plant in the Republic of Tanzania based on six exclusions and four decision criteria.

TABLE 10.1

Year-Wise Literature Survey of AHP MCDM Approach for Sustainable Energy Planning Decision Problems

Year	Author	Nationality	Research Purpose	Results
2009	Chatzimouratidis and Pilavachi (2009)	Greece	To identify the most suitable power plant for future energy generation covering economic, sustainable and technical factors by using AHP MCDM approach	They identified hydro power plant as the most suitable power plant, order followed by geothermal and wind energy
2010	Kahraman and Kaya (2010)	Turkey	To determine the most appropriate energy source in Turkey using fuzzy multi-criteria decision-making approach	They obtained that wind and solar energy are the most appropriate energy sources in Turkey for power generation application
2010	Daniel et al. (2010)	India	To prioritize the three most mature renewable energy sources in India using AHP MCDM approach	The prioritization order of renewable energy sources is wind, biomass and solar energy
2013	Stein (2013)	United States	To develop a multi-criteria decision-making approach for decision makers to rank different renewable and nonrenewable energy technologies	The ranking order of different energy sources: wind, solar, hydro, geothermal, gas, oil, nuclear, coal, biomass
2014	Ahmad and Tahar (2014)	Malaysia	To evaluate the most efficient renewable energy source in Malaysia for an application of electricity generation	They evaluated solar energy as the most efficient energy source in Malaysia for electricity generation
2014	Al-Qudaimi and Kumar (2018)	Malaysia	To develop a new intuitionistic fuzzy-AHP (IF-AHP) decision-making approach for the analysis of sustainable energy planning problems	They concluded that nuclear and solar energy are the suitable energy sources for the sustainable energy planning
2016	Al Garni et al. (2016)	Saudi Arabia	To prioritize renewable energy sources for sustainable electricity generation in the developed country of Saudi Arabia	The prioritization order follows; solar PV > solar thermal > wind > biomass > geothermal
2017	Haddad, Liazid and Ferreira (2017)	Algeria	To develop a multi-criteria decision-making approach to prioritize the renewable energy sources for the electricity generation system of Algeria	Solar energy is the most suitable option for the Algerian electricity system. Solar is followed by wind, geothermal, biomass and hydro power, respectively
2017	Jha and Puppala (2017)	India	To analyze the five available renewable energy sources and to rank them based on the Energy Index	The ranking order provides geothermal at the top followed by hydro, wind, biomass and solar energy
2018	Mirjat et al. (2018)	Pakistan	To assess the energy combination factors for the fulfillment of the long-term electricity supply of Pakistan	Energy efficiency and conversion are the best energy combination factor for the Pakistan electricity system
2018	Atabaki and Aryanpur (2018)	Iran	To propose a multi-criteria decision-making model for the identification of sustainable power source in Iran	Solar PV and combined cycle power plant are chosen as the sustainable power sources in Iran

10.3   METHODOLOGY OF ANALYTIC HIERARCHY PROCESS (AHP)

Thomas Saaty developed the AHP approach in 1980. AHP is most frequently used for research because it prepares a hierarchical or network-based structure of the given problem. Afterwards, it breaks into many subproblems, which are separately analyzed or solved (Štreimikiene, Šliogeriene and Turskis 2016). In the hierarchal structure, goal or objective represented is at the top level. Criteria and subcriteria are represented at the middle level and alternative represented at a lower level (Al Garni et al. 2016; Stein 2013). In the AHP approach, criteria and subcriteria were pair-wise compared to obtain relative importance. The alternatives are also pair-wise compared for the considered criteria and obtained for the importance index. A product of relative importance (weights) of criteria and category weight of alternatives was given the final weights for the ranking of the alternatives. In AHP, experts have to give relative importance/weightage within the range of a 1–9 scale (crisp value). As if, object A is equally important with object B then weight of 1 is given or if object A is strongly more important than B that means A is given weightage of 7 or B is assigned reciprocal of that weightage means 1/7 (Kahraman and Kaya 2010). The weight allotted in Saaty crisp scale (1–9) always contains some kind of impreciseness and vagueness (Tasri and Susilawati 2014). Therefore, the fuzzy linguistic scale was adopted over the Saaty crisp scale to overcome impreciseness and vagueness. The flow diagram of the research work is shown in the Figure 10.2.

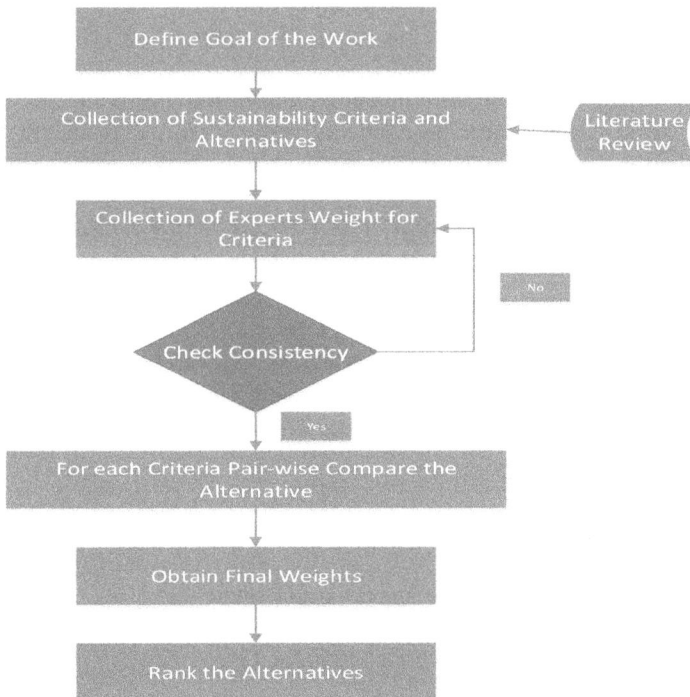

FIGURE 10.2   Flow diagram of the research work.

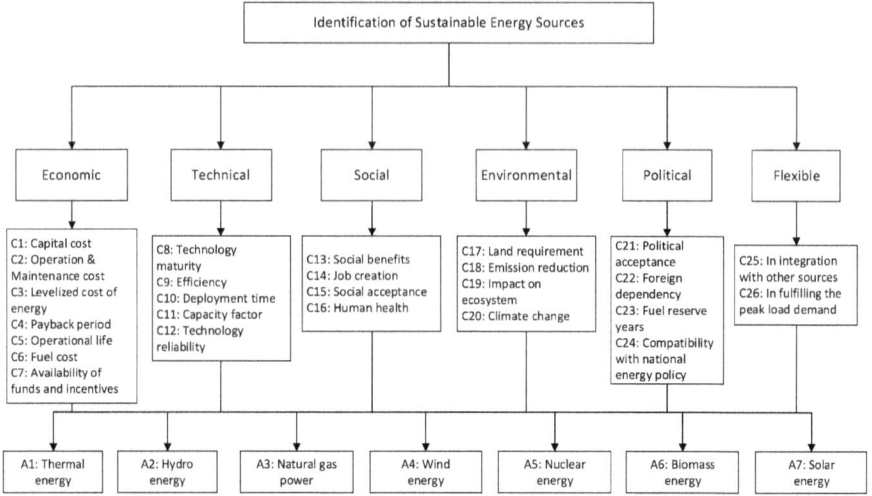

FIGURE 10.3   Hierarchical structure of sustainable energy decision problem.

Fuzzy AHP approach has the following steps.

1. Prepare the hierarchal network model to evaluate the most sustainable energy source in India as shown in Figure 10.3.
2. The research work employed the fuzzy AHP approach, which is proposed by the Buckley in 1985.
3. Perform pair-wise comparison among the considered criteria and subcriteria.
4. The relative importance (weights) is obtained using geometric mean method.

$$j = (p_1 \times p_2 \times \cdots \times p_n)^{1/n}, \quad k = (q_1 \times q_2 \times \cdots \times q_n)^{1/n},$$
$$l = (r_1 \times r_2 \times \cdots \times r_n)^{1/n} \tag{1}$$

n = number of criteria

5. Fuzzy weights are the product of fuzzy geometric mean (GM) and reciprocal of summation of that fuzzy geometric mean values as given in Eq. (2).

$$w_l = j_l (j_1 + j_2 + \cdots + j_n)^{-1}, \quad w_m = k_m (k_1 + k_2 + \cdots + k_n)^{-1},$$
$$w_u = l_u (l_1 + l_2 + \cdots + l_n)^{-1} \tag{2}$$

Where l = lower, m = middle and u = upper value

6. De-fuzzified crisp numeric value (DCNV) is the average of the fuzzy lower, middle and upper values as discussed in Eq. (3).

$$De - Fuzzified \ Crisp \ Numeric \ Value = \frac{w_l + w_m + w_u}{3} \tag{3}$$

7. The consistency ratio has been checked to validate the consistency of the given judgements. Consistency ratio should be less than 0.1 for true criteria weight.

$$Consistency\ Ratio = \frac{Cosistency\ Index}{Random\ Idex} \quad (4)$$

$$Consistency\ Index = (\lambda_{max} - n) / (n - 1) \quad (5)$$

Where, λ_{max} = maximum eigenvalue, n = number of elements

10.4 RESULT AND DISCUSSION

The decision maker uses the linguistic terminology to assign the relative importance to the considered criteria and alternatives. The linguistic terminology is converted into the crisp numeric values using a developed suitable scale. In this research work, there is a panel of three experts. The first expert is an academia with more than twenty years of research and teaching experience. The second expert is an environment professional who worked on several government and private environmental research projects. Third expert is an energy resource allocation professional with significant experience in the field of project allocation and implementation. First, the criteria were pair-wise compared using linguistic terminology. The linguistic weights have been converted into the fuzzy geometric mean by adopting the discussed methodology. Table 10.2 shows the fuzzy geometric mean, fuzzy weights and center of area of the considered criteria. The consistency ratio of the criteria weights table is 0.0208, which is less than 0.10. Therefore, the obtained weights are consistent.

Based on a pair-wise comparison among the criteria, economic criterion is chosen as a most effective criterion with a criteria weight of 0.194 as shown in Table 10.2. Environmental is the second effective criteria, followed by technical, political, social and flexible, as shown in Figure 10.4.

Afterwards, alternatives are pair-wise compared for environmental, technical, economic, social, political and flexible criteria. For each criterion, alternatives are assigned linguistic weightage and pair-wise compared. By pair-wise comparison of alternatives fuzzy geometric mean, fuzzy weights and de-fuzzified weights are obtained as presented in Tables 10.3–10.8.

TABLE 10.2
Determination of Criteria Weightage Using Fuzzy AHP MCDM Approach

Criteria	A Fuzzy Geometric Mean Value	Fuzzy Weights	COA (Center of Area)	Ranking
Economic	(1.026, 1.098, 1.380)	(0.147, 0.186, 0.268)	0.194	1
Technical	(0.920, 1.076, 1.215)	(0.132, 0.183, 0.236)	0.178	3
Social	(0.800, 0.987, 1.089)	(0.115, 0.168, 0.211)	0.160	5
Environmental	(0.987, 1.076, 1.328)	(0.142, 0.183, 0.257)	0.188	2
Political	(0.891, 0.981, 1.127)	(0.128, 0.167, 0.219)	0.167	4
Flexible	(0.531, 0.671, 0.826)	(0.076, 0.114, 0.160)	0.113	6

Consistency Ratio = 0.0208 < 0.10

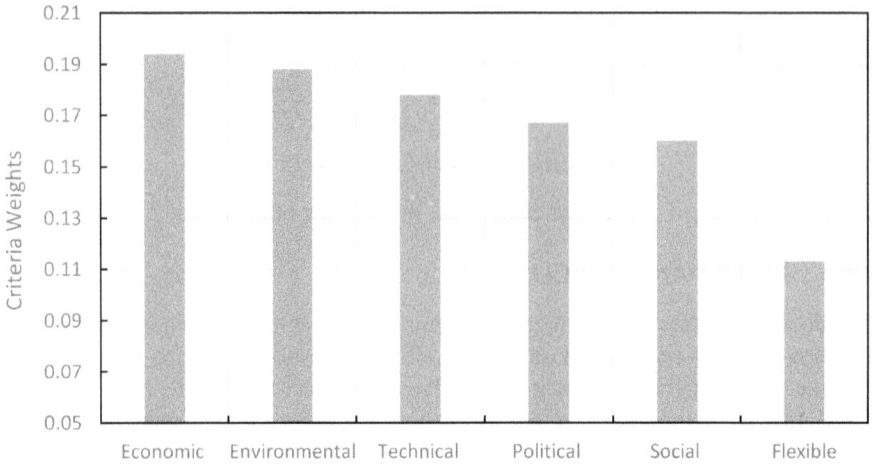

FIGURE 10.4   Criteria weights in decreasing order.

TABLE 10.3
Alternative Weights for Environmental Criterion

Environmental	Fuzzy Geometric Mean	Fuzzy Weights	De-Fuzzified Weights
Thermal	(0.503, 0.631, 0.829)	(0.056, 0.086, 0.137)	0.088
Hydro	(0.927, 1.156, 1.397)	(0.104, 0.157, 0.232)	0.156
Gas	(0.808, 0.938, 1.135)	(0.090, 0.127, 0.188)	0.128
Wind	(1.071, 1.324, 1.586)	(0.120, 0.180, 0.263)	0.178
Nuclear	(0.781, 0.921, 1.110)	(0.087, 0.125, 0.184)	0.126
Biomass	(0.683, 0.873, 1.031)	(0.076, 0.118, 0.171)	0.116
Solar	(1.258, 1.528, 1.862)	(0.141, 0.207, 0.309)	0.208

TABLE 10.4
Alternative Weights for Technical Criterion

Technical	Fuzzy Geometric Mean	Fuzzy Weights	De-Fuzzified Weights
Thermal	(0.705, 0.871, 1.025)	(0.083, 0.125, 0.180)	0.123
Hydro	(0.676, 0.770, 0.954)	(0.080, 0.110, 0.167)	0.113
Gas	(0.711, 0.861, 1.034)	(0.084, 0.123, 0.181)	0.123
Wind	(0.987, 1.223, 1.515)	(0.116, 0.175, 0.266)	0.176
Nuclear	(0.613, 0.802, 0.924)	(0.072, 0.115, 0.162)	0.110
Biomass	(0.836, 1.000, 1.222)	(0.098, 0.143, 0.214)	0.144
Solar	(1.173, 1.448, 1.820)	(0.138, 0.208, 0.319)	0.210

TABLE 10.5
Alternative Weights for Economic Criterion

Economic	Fuzzy Geometric Mean	Fuzzy Weights	De-Fuzzified Weights
Thermal	(0.898, 1.088, 1.297)	(0.103, 0.150, 0.215)	0.149
Hydro	(0.864, 1.011, 1.165)	(0.099, 0.139, 0.193)	0.137
Gas	(0.811, 0.885, 1.048)	(0.093, 0.122, 0.173)	0.124
Wind	(1.054, 1.289, 1.554)	(0.121, 0.178, 0.257)	0.177
Nuclear	(0.612, 0.749, 0.896)	(0.070, 0.103, 0.148)	0.102
Biomass	(0.6928, 0.841, 1.015)	(0.079, 0.116, 0.168)	0.116
Solar	(1.107, 1.3967, 1.771)	(0.127, 0.192, 0.293)	0.195

TABLE 10.6
Alternative Weights for Social Criterion

Social	Fuzzy Geometric Mean	Fuzzy Weights	De-Fuzzified Weights
Thermal	(0.602, 0.746, 0.926)	(0.066, 0.102, 0.159)	0.102
Hydro	(0.930, 1.194, 1.443)	(0.102, 0.163, 0.247)	0.160
Gas	(0.846, 1.051, 1.329)	(0.092, 0.143, 0.228)	0.145
Wind	(1.048, 1.342, 1.625)	(0.114, 0.183, 0.278)	0.180
Nuclear	(0.572, 0.708, 0.896)	(0.062, 0.097, 0.154)	0.097
Biomass	(0.717, 0.914, 1.189)	(0.078, 0.125, 0.204)	0.127
Solar	(1.123, 1.376, 1.753)	(0.123, 0.188, 0.300)	0.190

TABLE 10.7
Alternative Weights for Political Criterion

Political	Fuzzy Geometric Mean	Fuzzy Weights	De-Fuzzified Weights
Thermal	(0.543, 0.679, 0.838)	(0.060, 0.092, 0.142)	0.092
Hydro	(0.859, 1.101, 1.345)	(0.095, 0.150, 0.227)	0.148
Gas	(0.610, 0.751, 0.944)	(0.067, 0.102, 0.160)	0.103
Wind	(1.086, 1.307, 1.594)	(0.120, 0.178, 0.269)	0.178
Nuclear	(0.727, 0.914, 1.145)	(0.080, 0.124, 0.193)	0.125
Biomass	(0.928, 1.148, 1.402)	(0.103, 0.156, 0.237)	0.156
Solar	(1.165, 1.443, 1.782)	(0.129, 0.196, 0.301)	0.197

TABLE 10.8

Alternative Weights for Flexibility Criterion

Flexible	Fuzzy Geometric Mean	Fuzzy Weights	De-Fuzzified Weights
Thermal	(0.577, 0.740, 0.871)	(0.066, 0.102, 0.147)	0.100
Hydro	(1.100, 1.304, 1.627)	(0.125, 0.181, 0.275)	0.184
Gas	(1.048, 1.231, 1.523)	(0.120, 0.170, 0.257)	0.173
Wind	(0.835, 1.028, 1.276)	(0.095, 0.142, 0.215)	0.144
Nuclear	(0.670, 0.812, 1.008)	(0.076, 0.112, 0.170)	0.114
Biomass	(0.745, 0.936, 1.076)	(0.085, 0.130, 0.182)	0.126
Solar	(0.952, 1.170, 1.385)	(0.109, 0.162, 0.234)	0.160

In economic analysis, solar energy is chosen as a most economic feasible source by making a pair-wise comparison. Biomass and nuclear are chosen as the least cost-effective solution, as shown in Figure 10.5. Technically solar energy is chosen as a strong solution, followed by wind, biomass, gas, thermal, hydro and nuclear energy, respectively. In the environmental criteria, renewable energy sources (solar, wind and hydro) show the effective performance. Solar energy is chosen as the most socially accepted energy source and nuclear energy as the least accepted energy source in India. Politically renewable energy sources (solar, wind, biomass and hydro) are preferred to reduce foreign dependency and export of Indian currency. In the flexible criteria, hydro and gas energy sources show the best performances because of their capabilities to respond to peak load and easily on-off system.

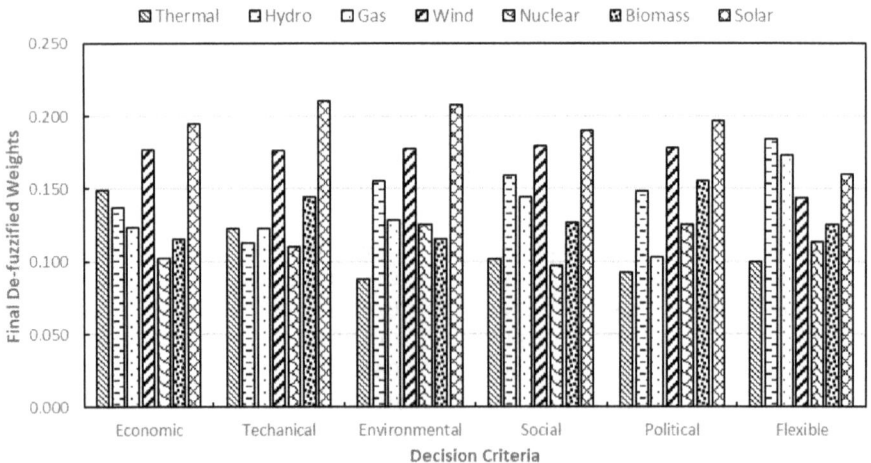

FIGURE 10.5 Representation of final defuzzified weight of each alternative in every criterion.

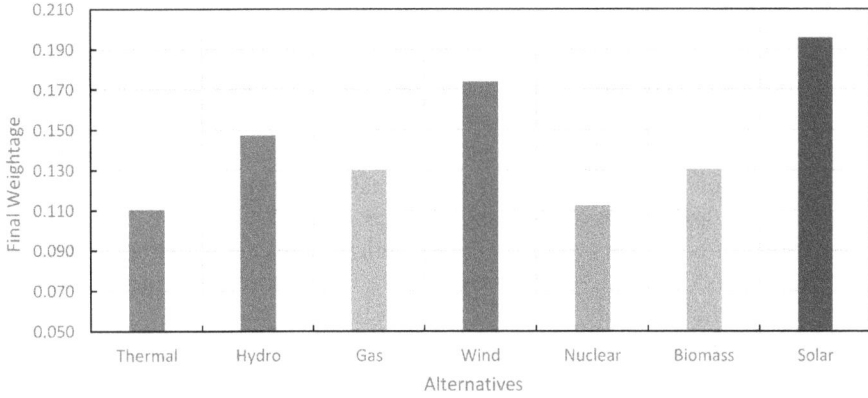

FIGURE 10.6 Representation of final weight of alternatives.

Final weight is obtained by the product of relative importance of criteria and alternatives. In the final weights, solar energy is at the first position with highest weight of 0.196. Wind energy is the second most sustainable energy, followed by hydro energy. Biomass and gas power combined shares the fourth position, with an equal weight of 0.130 as shown in Figure 10.6. Nuclear energy and thermal energy are the least sustainable energy sources, with the ranking position of fifth and sixth, respectively, as shown in Table 10.9.

TABLE 10.9
Determination of Final Weights and Ranking of Alternatives

	Economic	Technical	Environmental	Social	Political	Flexibility	Final Weightage	Final Ranking
Thermal	0.194 × 0.149	0.178 × 0.123	0.188 × 0.088	0.160 × 0.102	0.167 × 0.0924	0.113 × 0.100	0.110	6
Hydro	0.194 × 0.137	0.178 × 0.113	0.188 × 0.156	0.160 × 0.160	0.167 × 0.1483	0.113 × 0.184	0.147	3
Gas	0.194 × 0.124	0.178 × 0.123	0.188 × 0.128	0.160 × 0.145	0.167 × 0.1034	0.113 × 0.173	0.130	4
Wind	0.194 × 0.177	0.178 × 0.176	0.188 × 0.178	0.160 × 0.180	0.167 × 0.1782	0.113 × 0.144	0.174	2
Nuclear	0.194 × 0.102	0.178 × 0.110	0.188 × 0.126	0.160 × 0.097	0.167 × 0.1251	0.113 × 0.114	0.112	5
Biomass	0.194 × 0.116	0.178 × 0.144	0.188 × 0.116	0.160 × 0.127	0.167 × 0.1558	0.113 × 0.126	0.130	4
Solar	0.194 × 0.195	0.178 × 0.210	0.188 × 0.208	0.160 × 0.190	0.167 × 0.1968	0.113 × 0.160	0.196	1

10.5 CONCLUSIONS

Energy is an essential factor for the socio-economic development of societies and the nation. "Sustainable energy source" means economically viable, technically mature, socially accepted, politically motivated and environmentally friendly energy source. The Indian energy sector depends on both renewable and nonrenewable energy sources. In nonrenewable energy sources, thermal, gas power and nuclear cover the major portion, and in renewable energy sources, solar, wind, hydro and biomass cover the major portion in overall power generation. Each energy source has its own advantages and disadvantages, so there is a necessity to evaluate the most sustainable energy source. In the study, thermal (coal), nuclear, gas power, hydro, solar, wind and biomass, seven major energy generation sources, are considered. In the research work, six different criteria are considered – i.e., environmental, technical, economic, social, political and flexibility. Fuzzy AHP MCDM approach is employed for the criteria and alternative weights and rankings. Linguistic terminology is adopted for the expert judgements. By performing pair-wise comparison, economic criterion is selected as the most effective criterion. The effective criterion order is followed by environmental, technical, political, social and flexibility criterion. Solar energy shows the best performance among the considered criteria rather than flexibility criterion. In the final analysis, solar energy is chosen as a most sustainable energy source in India with final weight of 0.196. The second most sustainable energy source is wind energy, and hydro energy is the third most sustainable energy. Biomass and gas power combined at the fourth position. Nuclear energy is the fifth least sustainable energy source because of foreign dependency, export of Indian currency and less social acceptance. Thermal energy is the least sustainable energy source in India because of some serious issues of scarcity in the availability of the fuel, greenhouse gases emission and huge increase in the fuel price.

REFERENCES

Ahmad, Salman and Razman Mat Tahar. 2014. Selection of Renewable Energy Sources for Sustainable Development of Electricity Generation System Using Analytic Hierarchy Process: A Case of Malaysia. *Renewable Energy* 63: 458–66. Elsevier Ltd. Doi:10.1016/j.renene.2013.10.001.

Al-Qudaimi, Abdullah and Amit Kumar. 2018. Sustainable Energy Planning Decision Using the Intuitionistic Fuzzy Analytic Hierarchy Process: Choosing Energy Technology in Malaysia: Necessary Modifications. *International Journal of Sustainable Energy* 37(5): 436–37. Doi:10.1080/14786451.2017.1280496.

Aly, Ahmed, Steen Solvang Jensen, and Anders Branth Pedersen. 2017. Solar Power Potential of Tanzania: Identifying CSP and PV Hot Spots through a GIS Multicriteria Decision Making Analysis. *Renewable Energy* 113: 159–75. Elsevier Ltd. Doi:10.1016/j.renene.2017.05.077.

Amer, Muhammad and Tugrul U. Daim. 2011. Selection of Renewable Energy Technologies for a Developing County: A Case of Pakistan. *Energy for Sustainable Development* 15(4): 420–35. International Energy Initiative. Doi:10.1016/j.esd.2011.09.001.

Arce, María Elena, Ángeles Saavedra, José L. Míguez, and Enrique Granada. 2015. The Use of Grey-Based Methods in Multi-Criteria Decision Analysis for the Evaluation of Sustainable Energy Systems: A Review. *Renewable and Sustainable Energy Reviews* 47: 924–32. Elsevier. Doi:10.1016/j.rser.2015.03.010.

Atabaki, Mohammad Saeid and Vahid Aryanpur. 2018. Multi-Objective Optimization for Sustainable Development of the Power Sector: An Economic, Environmental, and Social Analysis of Iran. *Energy* 161: 493–507. Elsevier B.V. Doi:10.1016/j.energy.2018.07.149.

Baseer, M. A., S. Rehman, J. P. Meyer, and Md Mahbub Alam. 2017. GIS-Based Site Suitability Analysis for Wind Farm Development in Saudi Arabia. *Energy* 141: 1166–76. Elsevier B.V. Doi:10.1016/j.energy.2017.10.016.

Büyüközkan, Gülçin and Sezin Güleryüz. 2016. An Integrated DEMATEL-ANP Approach for Renewable Energy Resources Selection in Turkey. *International Journal of Production Economics* 182: 435–48. Elsevier. Doi:10.1016/j.ijpe.2016.09.015.

Chatzimouratidis, Athanasios I. and Petros A. Pilavachi. 2009. Sensitivity Analysis of Technological, Economic and Sustainability Evaluation of Power Plants Using the Analytic Hierarchy Process. *Energy Policy* 37(3): 788–98. Doi:10.1016/j.enpol.2008.11.021.

Çolak, Murat and İhsan Kaya. 2017. Prioritization of Renewable Energy Alternatives by Using an Integrated Fuzzy MCDM Model: A Real Case Application for Turkey. *Renewable and Sustainable Energy Reviews* 80(February): 840–53. Doi:10.1016/j.rser.2017.05.194.

Daniel, Joseph, Nandigana V R Vishal, Bensely Albert, and Iniyan Selvarsan. 2010. Multiple Criteria Decision Making for Sustainable Energy and Transportation Systems. *Lecture Notes in Economics and Mathematical Systems* 634: 13–27. Doi:10.1007/978-3-642-04045-0.

Energy Statistics. 2017."Ministry of Statistics and Programme Implementation". New Delhi. Available on: www.mospi.gov.in (Retrieved on 28 November 2019).

Garni, Hassan Al, Abdulrahman Kassem, Anjali Awasthi, Dragan Komljenovic, and Kamal Al-Haddad. 2016. A Multicriteria Decision Making Approach for Evaluating Renewable Power Generation Sources in Saudi Arabia. *Sustainable Energy Technologies and Assessments* 16: 137–50. Elsevier Ltd. Doi:10.1016/j.seta.2016.05.006.

Haddad, Brahim, Abdelkrim Liazid, and Paula Ferreira. 2017. A Multi-Criteria Approach to Rank Renewables for the Algerian Electricity System. *Renewable Energy* 107: 462–72. Elsevier Ltd. Doi:10.1016/j.renene.2017.01.035.

Hairat, Manish Kumar and Sajal Ghosh. 2017. 100 GW Solar Power in India by 2022 – A Critical *Review. Renewable and Sustainable Energy Reviews* 73(November 2016): 1041–50. Elsevier. Doi:10.1016/j.rser.2017.02.012.

Jha, Shibani K. and Harish Puppala. 2017. Prospects of Renewable Energy Sources in India: Prioritization of Alternative Sources in Terms of Energy Index. *Energy* 127: 116–27. Elsevier B.V. Doi:10.1016/j.energy.2017.03.110.

Kahraman, Cengiz and Ihsan Kaya. 2010. A Fuzzy Multicriteria Methodology for Selection among Energy Alternatives. *Expert Systems with Applications* 37(9): 6270–81. Doi:10.1016/j.eswa.2010.02.095.

Kahraman, Cengiz, Ihsan Kaya, and Selcuk Cebi. 2009. A Comparative Analysis for Multiattribute Selection among Renewable Energy Alternatives Using Fuzzy Axiomatic Design and Fuzzy Analytic Hierarchy Process. *Energy* 34(10): 1603–16. Elsevier Ltd. Doi:10.1016/j.energy.2009.07.008.

Kaya, Tolga and Cengiz Kahraman. 2010. Multicriteria Renewable Energy Planning Using an Integrated Fuzzy VIKOR & AHP Methodology: The Case of Istanbul. *Energy* 35(6): 2517–27. Elsevier. Doi:10.1016/j.energy.2010.02.051.

Klein, Sharon J.W. and Stephanie Whalley. 2015. Comparing the Sustainability of U.S. Electricity Options through Multi-Criteria Decision Analysis. *Energy Policy* 79: 127–49. Elsevier. Doi:10.1016/j.enpol.2015.01.007.

Kumar Shukla, Umesh and Seema Sharma. 2017. The Potential of Electricity Imports to Meet Future Electricity Requirements in India. *Electricity Journal* 30(3): 71–84. Elsevier. Doi:10.1016/j.tej.2017.03.007.

Lee, Hsing Chen and Ching Ter Chang. 2018. Comparative Analysis of MCDM Methods for Ranking Renewable Energy Sources in Taiwan. *Renewable and Sustainable Energy Reviews* 92(May): 883–96. Elsevier Ltd. Doi:10.1016/j.rser.2018.05.007.

Mainali, Brijesh, Shonali Pachauri, Narasimha D. Rao, and Semida Silveira. 2014. Assessing Rural Energy Sustainability in Developing Countries. *Energy for Sustainable Development* 19(1): 15–28. International Energy Initiative. Doi:10.1016/j.esd.2014.01.008.

Maxim, Alexandru. 2014. Sustainability Assessment of Electricity Generation Technologies Using Weighted Multi-Criteria Decision Analysis. *Energy Policy* 65: 284–97. Elsevier. Doi:10.1016/j.enpol.2013.09.059.

Ministry of Power. 2019. "Status of National Power." Available on: https://npp.gov.in/dash-Board/cp-map-dashboard (Retrieved on 16 November 2019).

Mirjat, Nayyar Hussain, Mohammad Aslam Uqaili, Khanji Harijan, Mohd Wazir Mustafa, Md Mizanur Rahman, and M. Waris Ali Khan. 2018. Multi-Criteria Analysis of Electricity Generation Scenarios for Sustainable Energy Planning in Pakistan. *Energies* 11(4): 1–33. Doi:10.3390/en11040757.

Pappas, Dimitrios and Konstantinos J. Chalvatzis. 2017. Energy and Industrial Growth in India: The Next Emissions Superpower? *Energy Procedia* 105: 3656–62. Elsevier. Doi:10.1016/j.egypro.2017.03.842.

Pohekar, S.D. and M. Ramachandran. 2004. Application of Multi-Criteria Decision Making to Sustainable Energy Planning – A Review. *Renewable and Sustainable Energy Reviews* 8(4): 365–81. Elsevier. Doi:10.1016/j.rser.2003.12.007.

Stein, Eric W. 2013. A Comprehensive Multi-Criteria Model to Rank Electric Energy Production Technologies. *Renewable and Sustainable Energy Reviews* 22: 640–54. Elsevier. Doi:10.1016/j.rser.2013.02.001.

Štreimikiene, Dalia, Jurate Šliogeriene, and Zenonas Turskis. 2016. Multi-Criteria Analysis of Electricity Generation Technologies in Lithuania. *Renewable Energy* 85: 148–56. Elsevier Ltd. Doi:10.1016/j.renene.2015.06.032.

Tasri, Adek and Anita Susilawati. 2014. Selection among Renewable Energy Alternatives Based on a Fuzzy Analytic Hierarchy Process in Indonesia. *Sustainable Energy Technologies and Assessments* 7: 34–44. Elsevier Ltd. Doi:10.1016/j.seta.2014.02.008.

Uyan, Mevlut. 2017. Optimal Site Selection for Solar Power Plants Using Multi-Criteria Evaluation: A Case Study from the Ayranci Region in Karaman, Turkey. *Clean Technologies and Environmental Policy* 19(9): 2231–44. Springer, Berlin, Heidelberg. Doi:10.1007/s10098-017-1405-2.

11 Fuzzy Based Evaluation of Optimal Service Time Using Execution/ Communication Time and Clustering

Anju Khandelwal
S.R.M.S. College of Engg. & Tech., Bareilly, India

Avanish Kumar
Commission for Scientific & Technical Terminology, MHRD
New Delhi, India

CONTENTS

11.1 INTRODUCTION

Various applications of assignment problem (AP) exist in the real world. Extensive examples can be seen in various areas of the assignment problem such as industry, commerce, technology, management science, etc. Real-life problems cannot be successfully solved by the classical optimal assignment problem. Fuzzy assignment problem (FAP) is more appropriate in today's context. The theory of fuzzy sets was given in 1965 by Lofty A. Jadeh. This theory represents impurity or uncertainty in daily life. With the increase in computer processing, the need for competencies in technology has increased even more rapidly. Expected handling effectiveness for special applications cannot be attained with a monotonous processing framework. These problems can be solved by distributed processing systems (DPS). The division and allocation of work are important in building DPS. If these steps are not executed properly, the structure of DPS increases the number of promoters that can reduce the overall flow of the structure. With advances in distributed systems, the assignment problem has turned into an important consideration. There is a need to improve efficiency for computing scheduling in broadcasting work. The most important topic when structuring any task algorithm is to minimize the extra time as much as possible.

An optimal solution is obtained by calculating a fuzzy mean service rate using the task allocation technique (Khandelwal, 2019) with the idea of fuzzy communication time. The author (Yadav et al., 2019) has allocated the processors efficiently on different nodes by balancing the load between them. In this process, care has also been taken to reduce execution and response time. An optimal solution to the TFA problem has been achieved using the centroid ranking technique (Mary and Selvi, 2018). In addition, it uses the Euclidean separation strategy to analyze the fuzzy value and its rank. Many researchers have solved the fuzzy assignment problem through a genetic algorithm approach. Through the genetic algorithm approach, each individual with cost (time) has been organized for only one task (Muruganandam and Hema, 2018) and is presented as an invariant number. In addition to the genetic algorithm approach here, crisp values have been derived by the Yager ranking method to obtain an optimal solution to the fuzzy assignment problem.

In various research papers, the defuzzification concept (Sharma, 2018, Kumar et al., 2013, Gani and Mohamed, 2013 and Yadav et al., 2011) is envisaged by the robust ranking method, wherein the fuzzy value is changed to one, i.e., crisp number one. In defuzzification, a mathematical model is formed to determine the correct response time of the system using triangular/trapezoidal fuzzy execution time and triangular/trapezoidal fuzzy intertask communication time. Fuzzy assignment problem is important in solving the real-life problem. In the real-life problem (Thakre et al., 2018, Selvi et al., 2017), the fuzzy assignment problem has been resolved through the example of four candidates/designations at Life Insurance Corporation (LIC). In this the solution of this problem is represented by the fuzzy triangular magnitude ranking method, Hungarian method, MOA and direct method. A diffusion strategy (Neelakantan and Sreekanth, 2016) has been proposed to accommodate tasks between different computers in the context of a problem. In this, the response time of assignment to a processor is limited from moving the overload computer to the low-load computer to using the load adjustment system from the computer. To assess the performance of the system, the load on the system is calculated by fluctuating the average differential arrival time of the tasks on each computer.

11.2   PROBLEM STATEMENT

Fuzzy Assignment Problem (FAP) is more appropriate for today's real-life problems. Here, we have a fuzzy assignment problem (FAP). The purpose of this fuzzy assignment problem is to allocate a set of processors $F_z\{p_j\}$ on a set of tasks $F_z\{t_i\}$ in such a way that the total impedance time is minimized. For the solution of the fuzzy assignment problem here we convert the fuzzy impedance time coefficients by using the fuzzy ranking method and then formed clusters of tasks by using K-means clustering technique.

Here, execution time, communication time, total response (impedance) time, defuzzification, and fuzzy mean service rate (Khandelwal, 2019) have been solved by using the following formulae.

- Fuzzy Execution Time: $F_z\{ET(\tilde{c})\} = \sum\limits_{1\le i\le m} \tilde{et}_{i,c(i)}$ (11.1)

- Fuzzy Communication Time:

$$F_z\{CT(\tilde{c})\} = \sum\limits_{\substack{1\le i\le m \\ i+1\le j\le m \\ i=j\neq k}} \{\tilde{ct}_{ik}\},\ k=1,2,3,\ldots,m;\ 1\le j\le n \qquad (11.2)$$

- Fuzzy Total Response Time:

$$F_z\{TRT(\tilde{c})\} = \max\limits_{1\le j\le n}\left(F_z\{ET(\tilde{c})\} + F_z\{CT(\tilde{c})\}\right) \qquad (11.3)$$

- Defuzzification: $R(\tilde{c}) = \dfrac{1}{2}\int\limits_{0}^{1}(\overline{a} + 3a_0 - \underline{a})f(k)dk$ (11.4)

- Clustering Technique: Clustering technique aims to partition m tasks into n cluster ($m \ge n$) in which each task belongs to the cluster with the adjacent mean. This technique produces exactly n different clusters of greatest possible discrepancy.

$$\textbf{\textit{Objective function } } Z_{min} = \sum\limits_{j=1}^{n}\sum\limits_{i=1}^{m}\left\|x_i^{(j)} - c_j\right\|^2 \qquad (11.5)$$

- Fuzzy Mean Service Rate: $F_z\{MSR(j)\} = \dfrac{\{Tt_j\}}{F_z\{ET(et_j)\}}$ (11.6)

11.3 COMPUTATIONAL ALGORITHM

The mapping between the tasks and processors is defined by $\varphi: N \to M$. A task may be a data file or code which is to be executed on different processors having different processing capabilities. Assume that number of tasks is more than the number of processors ($m \ge n$) as normally seen in real life. Also it is assumed that the execution time of a task on each processor and intertask communication time is known. The intertask communication time between the same tasks is zero.

Step 1: Set Fuzzy quantitative problem of m tasks $F_z\{t_i\}$ for $1\le i\le m$, i.e., $\{t_1,t_2,t_3,t_4,t_5\}$.
Step 2: Set Fuzzy quantitative problem of n processors $F_z\{p_j\}$ for $1\le j\le n$, i.e., $\{p_1,p_2,p_3\}$

Step 3: Set $F_z\{ET(et_{ij})\}$ and $F_z\{CT(ct_{ik})\}$ in the form of fuzzy triangular number. $F_z\{ET(et_{ij})\}$ and $F_z\{CT(ct_{ik})\}$ are taken in the form of matrices as Fuzzy Execution Time Matrix and Fuzzy Intertask Communication Time Matrix.

Step 4: Determine Defuzzified Crisp Value for $F_z\{ET(et_{ij})\}$ by using Eq. (11.4).

Step 5: Determine sum of each task processor-wise and stored in Fuzzy Sum Array $F_z\{ET_{Sum\_Row}\{\}\}$.

Step 6: Cluster the task into n processor.

Step 7: Select n points randomly as cluster centre and calculate the squared error function using Eq. (11.5).

Step 8: Assign task to their closest cluster centre.

Step 9: Calculate mean of all tasks in each cluster.

Step 10: Repeat steps 7, 8 and 9 until the same points are assigned to each cluster in consecutive rounds.

Step 11: Apply basic assignment method on these clusters so formed in step 10 and allocate these tasks clusters on different processors.

Step 12: Let tasks allocated to processors is denoted by $F_z\{CTET(et_{ij})\} = T_{allocate}$. Calculate Total Fuzzy Execution Time $F_z\{ET\}$, Total Fuzzy Communication Time $F_z\{CT\}$ and Total Fuzzy Task Response Time $F_z\{TRT\}$ using Eqs. (11.1), (11.2), and (11.3).

Step 13: Convert Total Task Response Time into crisp values once using step-4 because it is fuzzy quantitative number.

Step 14: Now calculate the Overall Task Response Time for all tasks allocated in all different processors and Fuzzy Mean Service Rate for each Processor $F_z\{MSR(,)\}$

Step 15: Stop.

11.4   PSEUDOCODE FOR METHOD

```
Begin

• Set F_z{t_i}where 1 ≤ i ≤ m and F_z denotes fuzzy allocation
  for task {t_i}
• Set F_z{p_j} where 1 ≤ j ≤ n
• Set Fuzzy Triangular Number F_z{ET(et_ij)} and F_z{CT(ct_ik)}
• Represent F_z{ET(et_ij)} and F_z{CT(ct_ik)} in rectangular matrix
  form (m > n)
• for i:=1 to m inclusive do

    for j:=1 to n inclusive do
        Calculate, R(c̃_ij) = 1/2 ∫₀¹ (ā + 3a₀ − a)f(k)dk
        F_z{ET(et_ij)} ← Mag(c̃_ij)
    end for
    end for
```

- Step5: $sum = 0$, $a =$

```
    for i:=1 to m inclusive do
    for j:=1 to n inclusive do
        if (((i % 2==0)|(i % 2==1)) && (j<=n)) then
            sum = et[i][j]+sum
                end if
                end for
            sum_a = sum
    sum = 0
            a = a + 1
    end for
```

- for i:=1 to m inclusive do
  ```
    for j:=1 to n inclusive do
      T(i, j) = (t_i % c_j)
      end for
      end for
    for i:=1 to m inclusive do
  ```
  $T(i)T(i)array)_{min}$
  $Cl_j = compare\ (diff(Tk_{k+1_{k+r}min},\ where\ k + r = j,\ 1 \le j \le n;$
  $r = 0, 1, 2, ...$
  ```
      then again
    ((C_k ← T(i)k + 1_{min_{min}}. ......|(C_{k+r} ← T(i)_{min}
      end for
  ```

- Calculate mean of all tasks $\{t_i\}$ in each cluster $\{Cl_j\}$
- Repeat steps 5, 6 and 7 until the same points are assigned to same cluster $\{Cl_j\}$ in consecutive rounds.
- Apply Hungarian method on these clusters so formed in step 8 and allocate
  $\{p_j\} \leftarrow \{t_i(Cl_j)\}$.
- Set $F_z\left\{Cl_{ET(et_{ij})}\right\} = T_{allocate} \leftarrow \{t_i(Cl_j)\}$
- Calculate Total Fuzzy Execution Time, $F_z\left\{ET(\tilde c)\right\} = \sum_{1 \le i \le m} \tilde et_{i,c(i)}$,

  Total Fuzzy Communication Time,

  $$F_z\left\{CT(\tilde c)\right\} = \sum_{\substack{1 \le i \le m \\ i+1 \le j \le m \\ i=j \ne k}} \{\tilde ct_{ik}\}\ ,\ k = 1, 2, 3, ..., m\ ;\ 1 \le j \le n$$

- Calculate overall FTRT for all tasks,
  $$F_z\left\{TRT(\tilde c)\right\} = \max_{1 \le j \le n}\left(F_z\left\{ET(\tilde c)\right\} + F_z\left\{CT(\tilde c)\right\}\right)$$

- Convert, $F_z\left\{TRT(\tilde c)\right\} \leftarrow \dfrac{1}{2}\displaystyle\int_0^1 (\bar a + 3a_0 - \underline a)f(k)dk$

- Calculate, $F_z\left\{MSR(j)\right\} = \dfrac{\{Tt_j\}}{F_z\{ET(et_j)\}}$

End

11.5 MATHEMATICAL IMPLEMENTATION

Steps 1–3: Set Fuzzy numerical problem for m tasks $F_z\{t_i\}$ for $1 \leq i \leq m$, i.e., $\{t_1, t_2, t_3, t_4, t_5\}$ and n processors $F_z\{p_j\}$ for $1 \leq j \leq n$, i.e., $\{p_1, p_2, p_3\}$ in matrix form. Also set fuzzy execution time (Table 11.1) $(F_z\{ET(et_{ij})\}/$ fuzzy communication time (Table 11.2) $F_z\{CT(ct_{ik})\}$ in the form of FTN in matrix form.

TABLE 11.1
Fuzzy Execution Time Matrix

$F_z\{ET(et_{ij})\} =$		t_1	t_2	t_3	t_4	t_5
	p_1	(5,10,20)	(10,15,20)	(10,20,30)	(5,10,20)	(5,10,15)
	p_2	(5,10,15)	(10,20,30)	(10,15,25)	(10,15,20)	(5,10,20)
	p_3	(10,15,20)	(10,15,25)	(10,15,20)	(5,10,15)	(5,15,20)

TABLE 11.2
Fuzzy Communication Time Matrix

$F_z\{CT(ct_{ik})\} =$		t_1	t_2	t_3	t_4	t_5
	t_1	(0,0,0)	17.5	10	31.25	6.25
	t_2	17.5	(0,0,0)	32.5	10	25
	t_3	10	32.5	(0,0,0)	10	10
	t_4	31.25	10	10	(0,0,0)	12.5
	t_5	6.25	25	10	12.5	(0,0,0)

Step 4: Using Ranking Method find Crisp Value for $F_z\{ET(et_{ij})\}$ obtained in Table 11.3.

TABLE 11.3
Defuzzified Crisp Values

	t_1	t_2	t_3	t_4	t_5
p_1	6.25	8.75	10	6.25	5
p_2	5	10	10	8.75	6.25
p_3	8.75	10	8.75	5	5

Step 5: Determine sum of each task (processorwise) and stored in Fuzzy Sum Array $F_z \{ET_{Sum\_Row}\}$ shown in Table 11.4.

TABLE 11.4
Fuzzy Task_Sum Array

$F_z \{ET_{sum\_row}(,)\} =$	Task	t_1	t_2	t_3	t_3	t_5
	Crisp Value	20	28.75	28.75	20	16.25

Step 6: Cluster the task into three processors.
Steps 7–10: Select three points $\{18, 22, 26\}$ randomly as cluster centre and calculate the squared error function (by Eq. [11.5]). Tables 11.5–11.7 representing the stepwise clustering iterations.

TABLE 11.5
Centred Mean Value in Three Different Iterations

	c_1	c_2	c_3
$t_1 = 20$	18 / 18.125 / 16.25	22 / 20_{2,3}	26 / 28.75_{2,3}
$t_2 = 28.75$	18 / 18.125 / 16.25	22 / 20_{2,3}	26 / 28.75_{2,3}
$t_3 = 28.75$	18 / 18.125 / 16.25	22 / 20_{2,3}	26 / 28.75_{2,3}
$t_4 = 20$	18 / 18.125 / 16.25	22 / 20_{2,3}	26 / 28.75_{2,3}
$t_5 = 16.25$	18 / 18.125 / 16.25	22 / 20_{2,3}	26 / 28.75_{2,3}

TABLE 11.6
Deviation During Clustering in Three Different Iterations

	d_1	d_2	d_3
$t_1 = 20$	2 / 1.875 / 3.75	2 / 0_{2,3}	6 / 8.75_{2,3}
$t_2 = 28.75$	10.75 / 10.625 / 12.25	6.75 / 8.75_{2,3}	2.75 / 0_{2,3}
$t_3 = 28.75$	10.75 / 10.625 / 12.25	6.75 / 8.75_{2,3}	2.75 / 0_{2,3}
$t_4 = 20$	2 / 1.875 / 3.75	2 / 0_{2,3}	6 / 8.75_{2,3}
$t_5 = 16.25$	1.75 / 1.875 / 0	5.75 / 3.75_{2,3}	9.75 / 12.5_{2,3}

TABLE 11.7

Nearest Cluster with New Centroid Value in Different Iterations

	Nearest Cluster	NewCentroid
$t_1 = 20$	× / 2 / 2	20 / 20* / 20*
$t_2 = 28.75$	3 / 3 / 3	28.75* / 28.75* / 28.75*
$t_3 = 28.75$	3 / 3 / 3	* / * / *
$t_4 = 20$	3 / 3 / 3	18.125 /* /* /
$t_5 = 16.25$	1 / 1 / 1	18.125* /* /16.25

In 2nd and 3rd iteration same points are assigned to each cluster. Hence tasks clustered on three processors as $\{t_1 \oplus t_4\}, \{t_2 \oplus t_3\}\{t_5\}$ in three consecutive rounds.

Step 11: Apply basic Assignment procedure to allocate clustered tasks on the different processors. It is shown in Table 11.8.

TABLE 11.8

Assignment Method on Cluster

	$t_1 \oplus t_4$	$t_2 \oplus t_3$	t_5
p_1	12.5	18.75	5
$F_z\{Cl_{ET(et_{ij})}\} = T_{allocate} \Rightarrow$ p_2	13.75	20	6.25
p_3	13.75	18.75	5

Steps 12–14: Evaluate $F_z\{ET\}$, $F_z\{CT\}$ and $F_z\{TRT\}$ using Eqs. (11.1), (11.2), and (11.3). Also Evaluate Fuzzy $F_z\{MSR(,)\}$ for each Processor. These values are shown in Table 11.9.

Step 15: Stop

TABLE 11.9

Fuzzy Total Response Time and Mean Service Rate

Processor	Tasks	$F_z\{ET\}$	$F_z\{CT\}$	$F_z\{TRT\}$	CrispValues	$F_z\{MSR\}$
p_1	$t_1 \oplus t_4$	12.5	(70,120,175)	(80,140,215)	78.5	0.0254
p_2	t_5	6.25	(60,95,130)	(65,105,145)	58.75	0.0170
p_3	$t_2 \oplus t_3$	18.75	(90,145,205)	(110,180,255)	101.25	0.0197

11.6   CONCLUSION AND DISCUSSION

In this research problem, we have established a fuzzy system using fuzzy execution time and fuzzy intertask communication time through which the optimal value of the mechanism can be obtained. Here, tasks are moved from overload processors to underload processors using a mixture of functions to reduce the average service rate of tasks. To estimate the optimal value of the system according to the process, different tasks have been allocated according to the average service rate of the tasks on each processor. Compared to existing approaches, the proposed technique has yielded the optimal value minimum. The proposed approach reduces the average service rate of the system for an effective mix of tasks and optimal allocation, which is an important part of this technique. According to the approach proposed here the optimum value is 101.25 which is lower than the 66.83 and 65.16 values as compared to the existing approaches.

The methodology proposed in this chapter qualifies for the allocation of the processor's functions and provides optimal values, but we cannot say that this currently existing technique provides perfection to this mechanism. We can use various other methods of solving various technology in this field. Here, we have only focused on reducing the average service rate of the model, but this may also apply to load balancing problems with different models of phased technology in the future.

REFERENCES

Gani A. N. and V. N. Mohamed. 2013. Solution of a Fuzzy Assignment Problem by Using a New Ranking Method, *International Journal of Fuzzy Mathematical Archive*, 2, 8–16, ISSN: 2320 –3242 (P), 2320 –3250 (online).

Khandelwal A. 2019. Fuzzy Based Amalgamated Technique for Optimal Service Time in Distributed Computing System, *International Journal of Recent Technology and Engineering*, 8(3), 6763–6768, ISSN: 2277-3878.

Kumar H., M. P. Singh and Pradeep Kumar Yadav. 2013. A Tasks Allocation Model with Fuzzy Execution and Fuzzy Inter-Tasks Communication Times in a Distributed Computing System, *International Journal of Computer Applications*, 72(12), 24–31, ISSN: 0975-8887.

Mary R. Q. and D. Selvi. 2018. Solving Fuzzy Assignment Problem Using Centroid Ranking Method, *International Journal of Mathematics and its Applications*, 6(3), 9–16, ISSN: 2347-1557.

Muruganandam S. and K. Hema. 2018. A Method of Solution to Fuzzy Assignment Problem Using Genetic Algorithm, *Global Journal for Research Analysis*, 7(4), 154–157, https://www.doi.org/10.36106/gjra, ISSN: 2277-8160.

Neelakantan P. and S. Sreekanth. 2016. Task Allocation in Distributed Systems, *Indian Journal of Science and Technology*, 9(31), DOI: 10.17485/ijst/2016/v9i31/89615.

Selvi D., R. Queen Mary and G. Velammal. 2017. Method For Solving Fuzzy Assignment Problem Using Magnitude Ranking Technique, *International Journal of Applied and Advanced Scientific Research*, Special Issue, 16–20, ISSN: 2456-3080.

Sharma A. 2018. Time Optimization Model with Defuzzification for the Task Allocation in Distributed Computing System, *International Journal of Electronics Engineering*, 10(2), 492–501, ISSN: 0973-7383.

Thakre T. A., Onkar K Chaudhari and Nita R Dhawade. 2018. Placement of Staff in LIC Using Fuzzy Assignment Problem, *International Journal of Mathematics Trends and Technology (IJMTT)*, 53(4), 259–266, ISSN: 2231-5373.

Yadav P. K., P. Pradhan and Preet Pal Singh. 2011. A Fuzzy Clustering Method to Minimize the Inter Task Communication Effect for Optimal Utilization of Processor's Capacity in Distributed Real Time Systems, Proceedings of the International Conference on SocProS. K. Deep et al. (Eds.). springerlink.com, AISC 130, 159–168.

Yadav S., Rakesh Mohan and P. K. Yadav 2019. Fuzzy Based Task Allocation Technique in Distributed Computing System, *International Journal of Information Technology*, 11(1), 13–20, https://doi.org/10.1007/s41870-018-0172-6.

12 Solving Knapsack Problem with Genetic Algorithm Approach

Manish Saraswat
Geetanjali Institute of Technical Studies, Udaipur, India

R. C. Tripathi
Teerthanker Mahaveer University, Moradabad, India

CONTENTS

12.1 CONCEPT OF GENETIC ALGORITHMS

Genetic algorithms (GA) are classical, heuristic search algorithms that conceptually use the approaches that are adopted by living organisms to survive in their environment. The genetic algorithms are Socratic in nature, their performance and output efficiency can vary across multiple executions [1]. In genetic algorithms, solutions are interpreted as optimization problem. The genetic algorithms work on population and coding of parameter sets, which are required for solution. The population is constructed by sequent of numbers called "chromosomes" and a sequence of number that constructs the chromosome is named as "gene". The assemblage of various chromosomes is called a population. The initial population is generated randomly without any specific pattern. Each chromosome in a population has a fitness value, which is a mathematical expression and determines the probability of survival of

Step 1: Log in the population randomly on the basis of program and credits.

Step 2: Evaluate the fitness of individual chromosome and choose the chromosome which have higher fitness for construction of next generation offspring.

Step 3: Execute the Crossover operator with high probability at appropriate points to exchange bit values.

Step 4: Place the Mutation operator with low probability to mutate or change the bit value in one chromosome.

Step 5: Re-calculate the fitness of new generated population and insert the new population (set new population as current population).

Step 6: If criteria of termination met, stop the search otherwise go to step-2

FIGURE 12.1   Six essential steps for genetic algorithm implementation.

each particular chromosome in the next generation. The outlines of GA are shown as in Figure 12.1 [2].

12.2   THE KNAPSACK PROBLEM

The knapsack problem (KP) is a specimen of stochastic and combinatorial optimization problem [3] that uses to find the optimum solution among available solutions [4]. Mathematically, it can be formulated by numbering the object 1 to n and introducing a vector for binary variables $x_j (j - 1...n)$ having the meaning [5]:

$$x_{j=} \begin{cases} 1 \\ 0 \end{cases} \text{ is '1', if object selected, otherwise '0';}$$

In this chapter, we will focus on 0/1 KP, in which 'n' items (objects) will be selected for a knapsack having capacity of 28 kg and it is assumed a knapsack has positive integer weight [2].

object/Item 'i' has a positive integer volume 'W_i' and positive integer benefit 'B_i'. In addition, there are 'Q_i' copies of item 'i' available, where quantity 'Q_i' is a positive integer satisfying $1 \le Q_i \le \infty$.

Let's 'X_i' determines how many copies of item 'i' can be placed into the knapsack. The goal is to:

$$\text{Maximize} \sum_{i=1}^{N} B_i X_i$$

Subject to the constraints $\sum_{i=1}^{N} W_i X_i \leq W$

and $0 \leq X_i \leq Q_i$

The KP is bounded and may be either 0 or 1 or multi-constraints. If one or more of the Q_i is infinite, the KP is unbounded; otherwise, the KP is bounded [6]. If $Q_I = 1$ for $i = 1, 2, N$, the problem is a 0-1 knapsack problem. In this paper the problem is 0/1, therefore we worked on bounded 0-1 KP, we could not have more than one copy of an item in the knapsack.

12.3 STEPS INVOLVED IN GENETIC ALGORITHM

Let's assume, we are going to spend a month in an uncultivated region. Suppose we have knapsack (or carry bag) that has maximum capacity of **28 kg**. Here different useful items and each having its own "benefit points" and "weight". In this problem our objective is to maximizing the beneficial (B_i) point, so that a carrying bag has maximum useful items. Detail of each item is given in Table 12.1.

12.3.1 INITIALIZATION

Providing the solution of a given bounded knapsack problem using genetic algorithm, first create population, it has individuals and each individual has their own set of chromosomes. In this problem, the genotype structure of chromosomes is "binary" strings. If structure has "1", it means that item is considered, and if "0", it means item is not considered or dropped as shown in Figure 12.2.

12.3.2 FITNESS FUNCTION

To calculate the fitness of all chromosomes, B_i points are considered as survival of chromosomes. The overall survival points are shown in the Table 12.2.

TABLE 12.1
Showing Items with Benefit and Weight

S. No.	Items	Benefit (Bi)	Weight (Wi)
1	Sleeping bag	14	16
2	Rope	4	6
3	Pocket knife	5	8
4	Torch	6	8
5	Bottle	9	7
6	Glucose	15	6

Chromosome-1					
1	0	0	1	1	0

Collection of
Chromosomes
Known as Population

Chromosome-2
Gene-1

0	0	1	1	1	0

Chromosome-3
Gene-2

0	1	0	1	0	0

Chromosome-4
Gene-3

0	1	1	0	0	1

Gene-4

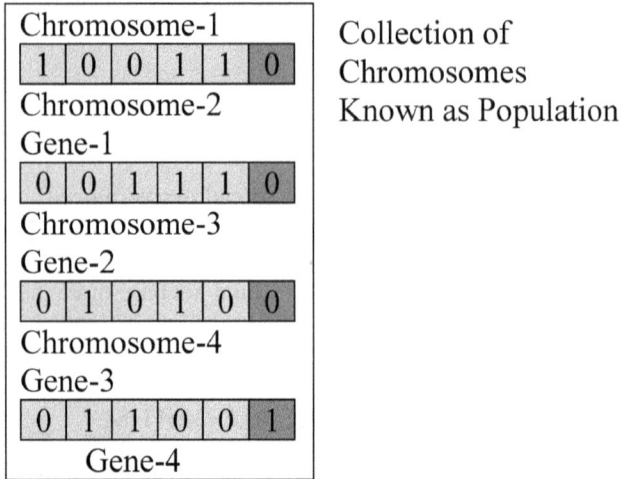

FIGURE 12.2 Relations between gene, chromosome and population with four chromosomes, and each chromosome has six values depending inclusion or exclusion of genes.

Now, it is clear chromosomes have different survival points and the fitness order would be as follows:

Chromosome-1>chromosome-4>chromosome-2> chromosome-3

12.3.3 SELECTION

After calculating the fitness of each chromosome, the fittest are selected from the pool, which come together to produce their offspring. Generally, we should reproduce the chromosomes that have more fitness value. Although, it creates convergence of population as less diversity in solutions within few generations. To maintain the diversity between chromosomes, we generally use roulette wheel selection method. In this method, selection is based on the statistical value of individual. The individuals showing higher fitness value on the wheel (Figure 12.3) get more chances to create a new population, while individuals showing poor fitness are rejected. Now

TABLE 12.2

Calculation of Fitness Points of Chromosomes on the Basis of Chromosome Inclusion/Exclusion

Chromosome	Survival Points
Chromosome-1[100110]	[14+0+0+6+9+0=29]
Chromosome-2[011100]	[0+4+5+6+0+0=15]
Chromosome-3[010100]	[0+4+0+6+0+0=10]
Chromosome-4[011001]	[0+4+5+0+0+19=28]

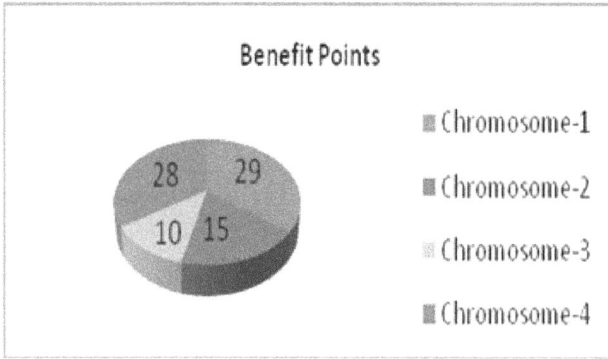

FIGURE 12.3 Four sections of chart showing benefits points of various chromosomes and chromosome-1 has maximum value while chromosome-3 has minimum.

consider the problem where population of chromosomes is divided in "m" number of divisions. The plot area engrossed by each chromosome is equivalent to its fitness value. Based on chromosome fitness values as in Table 12.2, fitness of chromosomes can be seen in Figure 12.3.

When the rotation of the wheel starts, the area that comes in front of the fixed point is chosen as the parent and is repeated exactly for the second parent.

12.3.4 CROSSOVER

Now crossover operation is applied on chosen parent chromosomes to generate two new members for the next generation to produce offspring. Let us find the crossover between chromosome-1 and chromosome-4, which were chosen in previous step on the basis of their fitness value. It is shown in Figure 12.4.

The crossover shown in the Figure 12.4 is most fundamental, and it is known as a one-point crossover. In this, random crossover points are identified and tails of both chromosomes exchanged to produce new offspring. Now if crossover operation performs on two points, it known as multipoint crossover. Crossover is important to maintain diversity between individuals [7].

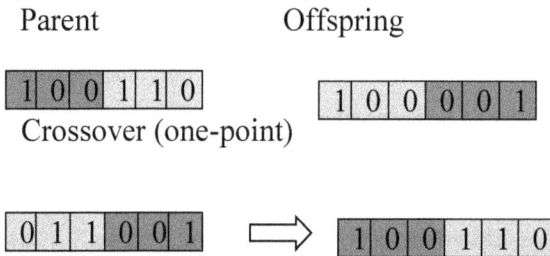

FIGURE 12.4 Crossover operation between chromosome-1 and chromosome-4 (one point) and resultant create offspring.

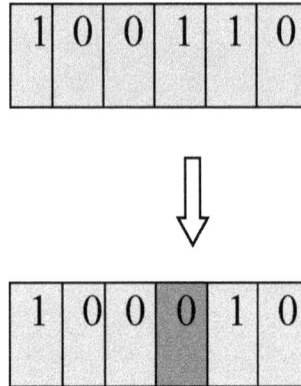

FIGURE 12.5   Mutation operation where genes are flipped: 1 replaced by 0.

12.3.5   MUTATION

Change in chromosome structure or randomly alteration in genes is known as "mutation"; therefore, a random swapping of gene on chromosome is done here, which promotes the idea of diversity in the population [6]. Process of mutation is illustrated in the Figure 12.5.

12.4   TERMINATION CONDITION

The experiment terminates either the specified number of generations achieved or no improvement seen for 'x' iterations. If fitness function attains prespecified value, the experiment winds up then too.

12.4.1   STEPS FOR KNAPSACK PROBLEM SOLVING WITH GA

Genetic algorithm is best known in design; here genotype of chromosomes is represented by binary combination like "10101011". For the stated problem of optimization in an experiment, an initial population of binary strings is created randomly and quality of each chromosome is determined by the value of fitness function. The fitness value/fitness function is calculated as:

> If $\left(\text{knapsack capacity}\right) >= $ Population's individuals total weight
>
> Then fitness value = population's individuals total weight
>
> Else Fitness = 0

Once population is created, parents are selected from the pool. The crossover operation is applied to create "offspring" and fitness of each chromosome is evaluated. In the next step, mutation operation is performed to maintain more diversity in population and result [4]. This allows GA to analyze more search space, by this it avoid

TABLE 12.3
List of Items with Symbols (X₁...X₆) in Chromosomes Structure and if Item Included the Gene Value is 1 Otherwise 0

S. No.	Item	Symbol	Values
1	Sleeping bag	X_1	If included in combination it is '1'
2	Rope	X_2	otherwise it is '0'
3	Pocket knife	X_3	
4	Torch	X_4	
5	Bottle	X_5	
6	Glucose	X_6	

local optima. For our knapsack problem, we consider the data with $n = 6$ and $W = 28$. The chromosomes based on item inclusion or exclusion are represented in Table 12.3.

In our case, the problem consists of population size $n = 6$ and the solution is focus on of selecting appropriate item $(X_1, X_2,...Xn)$ in the knapsack. The chromosome represents the vector X, consist the value "1" if item is included "0" if item is not included. It is shown as:

$$\boxed{X_1 \quad X_2 \quad X_3 \quad .... \quad Xn}$$

We have created the population by taking the chromosomes values randomly; a few combinations of potential solutions may be as follows:

> 000001,100001,100100,110100,001010,011000,111111,001110,
> 110001,111000,110000,101000,1101011,010101,100010............

Ten percent (say population size 100) of chromosomes are selected from the population to perform crossover operation for creating offspring and mutation operation to maintain the diversity of population. We select the chromosomes (items) by computing R_i to benefit to weight. The chromosomes (item) having more benefit and low weight for the knapsack are selected. The evaluation process remains continue to N generations, for a first generation is shown in Table 12.4.

This process continues until an optimized solution is obtained or termination criteria is met. GA can be implemented by following steps:

```
Import numpy # Version 3.8
# Define data
# Define Benefits
p <- c(4,14,5,6,9,15)
#Define Weight
w <- c(16,6,8,8,7,6)
#Define Knapsack Capacity (W)
W <- 28
```

TABLE 12.4

Estimation of Benefit (B$_i$) and Weight (W$_i$) Ratio (R$_i$) for First Generation and Same Process will be Continue Till Termination

S. No.	Item (Chromosome)	Benefit (B$_i$)	Weight (W$_i$)	R$_i$ = (B$_i$/W$_i$)
1	000001	15	6	2.50
2	100001	29	22	1.318182
3	100100	23	24	0.958333
4	110100	24	30	0.80
5	001010	14	15	0.933333
6	110000	18	22	0.818182
7	111111	53	51	1.039216
8	001110	20	23	0.869565
9	110001	33	28	1.178571
10	111000	23	30	0.766667

```
# Define Number of Objects (Items)
n <- length (p)
# Define Fitness Function
knapsack <- function(x)
f <- sum(x * p)
penalty <- sum(w) * abs(sum(x * w) - W)
f - penalty
ga(type="binary",
Fitness=knapsack,
N-Bits=n,
# Set Maximum Number of Generations
# Stopping Criteria (run) if best value obtained or no
improvements in run generations
maxiter=100,
run=150,
pop-size=100,
seed=101)
x.star<- SGA@ solution# Final solution: c(1,1,0,0,0,1)
sum(x.star) # Total number of selected items: 3
sum(x.star * p) # Total profit/benefit of selected items: 33
sum(x.star * w) # Total weight of selected items: 30
```

12.5　CONCLUSION

To solve our problem, the solution found by GA within knapsack's capacity constraint of $W = 28$ is $\mathbf{x}^* = (1,1,0,0,0,1)$. It is observed that the optimal solution for selecting items is {1, 2, 6}, which is maximum total benefit of $f(\mathbf{x}^*) = 33$. As the outcome of this experiment, GAs are meta-algorithms that can apply for obtaining solutions of optimizing problems. Here, we have discussed the application of GA to one instance of constrained optimization: 'knapsack problem'. In this experiment, it is demonstrated that GA can be applied to yield a higher benefit or total profit.

REFERENCES

1. Zhou, Y., Bao, Z., Luo, Q. and Zhang, S. 2017. A complex-valued encoding wind driven optimization for the 0-1 knapsack problem. *Applied Intelligence*. 46: 684–702.
2. Yang, X.-S. 2017. Nature-Inspired Algorithms and Applied Optimization, Springer, New York, NY.
3. Rezoug, A., Bader-El-Den, M. and Boughaci, D. 2018. Guided genetic algorithm for multidimensional knapsack problem. *Memetic Computing*. 10(1): 29–42.
4. Yu, X. and Gen, M. 2010. Introduction to Evolutionary Algorithms, Decision Engineering Series, Springer, New York, NY.
5. Martello, S. and Toth, P. 1990. Knapsack Problems, DEIS University, Bologna, Italy.
6. Scrucca, L. 2013. GA: A package for genetic algorithms in R. *Journal of Statistical Software*. 53(4): 1–37.
7. Feng, Y.H. and Wang, G.-G. 2018. Binary moth search algorithm for discounted {0,1} knapsack problem. *IEEE Access*. 6: 10708–10719.

Section III

Statistical Models

13 A Difference-Cum-Exponential Type Efficient Estimator of Population Mean

Kuldeep Kumar Tiwari, Sandeep Bhougal
School of Mathematics, Shri Mata Vaishno Devi University
Katra, India

Sunil Kumar
University of Jammu, Jammu and Kashmir, India

CONTENTS

13.1 INTRODUCTION

Precision is the reason for most of the research work in statistical theory and computation. Auxiliary information usually enhances the precision in survey sampling. It can be used at the selection stage or in estimation or on both. In the literature, auxiliary information is used by authors to construct various efficient estimators of different population parameters. The ratio estimator works better when there is a positive correlation between study and auxiliary variables, while the product estimator performs well in case of the negative correlation. Generally, a linear regression estimator is better than both ratio and product estimators. The equality in the efficiencies is obtained if the intercept of the regression line of y on x is zero. Since the regression estimator either performs better or equal to ratio and product estimators, it is convenient to use linear regression estimator than classical ratio and product estimator to gain in the efficiency. Many authors, such as [1–16], etc., worked on ratio and product type estimator to get the better results of such type estimator. Further, to get a better result, researchers worked on difference and exponential type estimators and showed its usefulness. Some of them are [17–20], etc.

Motivated by Rao [21] and Ekpenyong and Enang [3], in this paper we propose a difference-cum-exponential type estimator and studied their properties. In section 13.2, we made a review of the existing estimator related to the proposed estimator. Later, we suggest a difference cum exponential type estimator and derive the expression for the bias and mean squared error and given the conditions for minimum mean square error. Last, we do practical analysis by a simulation study and real-life populations to illustrate the performance of the proposed estimator.

13.2　NOTATIONS AND LITERATURE REVIEW

Let a finite population $\cup = \{U_1, U_2, .., U_N\}$ of size N. A sample of size n is taken under simple random sampling without replacement technique. Let y be the study variable and x be the auxiliary variable. y_i and x_i, $i = 1, 2, .., N$ be the observations of y and x at i^{th} unit in the population. We will use the following notations in the subsequent work.

Sample mean for study variable $\bar{y} = \dfrac{1}{n} \sum_{i=1}^{n} y_i$, sample mean for auxiliary variable $\bar{x} = \dfrac{1}{n} \sum_{i=1}^{n} x_i$, population mean for study variable $\bar{Y} = \dfrac{1}{N} \sum_{i=1}^{N} y_i$, population mean for auxiliary variable $\bar{X} = \dfrac{1}{N} \sum_{i=1}^{N} x_i$, population ratio $R = \dfrac{\bar{Y}}{\bar{X}}$, population variance for y, $S_y^2 = \dfrac{1}{N-1} \sum_{i=1}^{N} (y_i - \bar{Y})^2$, population variance for x, $S_x^2 = \dfrac{1}{N-1} \sum_{i=1}^{N} (x_i - \bar{X})^2$, population covariance $S_{xy} = \dfrac{1}{N-1} \sum_{i=1}^{N} (y_i - \bar{Y})(x_i - \bar{X})$, $C_y = \dfrac{S_y}{\bar{Y}}$, $C_x = \dfrac{S_x}{\bar{X}}$, $\rho = \dfrac{S_{xy}}{S_x S_y}$, $\beta_{yx} = \dfrac{S_{xy}}{S_x^2}$, $C_{xy} = \rho C_x C_y$, $K_{yx} = \dfrac{\beta_{yx}}{R}$, $f = \dfrac{n}{N}$, $\lambda = \dfrac{1-f}{n}$.

Some estimators of population mean and their mean square error (MSE) from literature are presented here.

Sample mean $\bar{y}$ is the usual unbiased estimator of population mean with variance,

$$\text{var}(\bar{y}) = \lambda \bar{Y}^2 C_y^2 \tag{13.1}$$

The classical ratio estimator to estimate the population mean $\bar{Y}$ when the population mean of auxiliary variable x is known, defined as

$$T_R = \bar{y} \left(\dfrac{\bar{X}}{\bar{x}} \right) \tag{13.2}$$

The *MSE* of the classical ratio estimator is

$$MSE(T_R) = \lambda \bar{Y}^2 [C_y^2 + C_x^2(1 - 2K_{yx})] \tag{13.3}$$

The expression for linear regression estimator is written as

$$T_{RE} = \bar{y} + \beta_{yx}(\bar{X} - \bar{x}) \tag{13.4}$$

and MSE of regression estimator is given as

$$MSE(T_{RE}) = \lambda \bar{Y}^2 C_y^2(1 - \rho^2) \tag{13.5}$$

In 1991, Bahl and Tuteja [17] suggested the ratio type exponential estimator as

$$T_{expR} = \bar{y} \quad \exp\left(\frac{\bar{X} - \bar{x}}{\bar{X} + \bar{x}}\right) \tag{13.6}$$

and obtained MSE as

$$MSE(T_{expR}) = \lambda \bar{Y}^2 \left[C_y^2 + \frac{1}{4}C_x^2(1 - 4K_{yx}) \right] \tag{13.7}$$

Rao [21] proposed a difference estimator as

$$T_{RAO} = h_1\bar{y} + h_2(\bar{X} - \bar{x}) \tag{13.8}$$

where h_1 and h_2 are suitable constants.
    MSE of the defined estimator for optimum values of h_1 and h_2 is obtained as

$$MSE(T_{RAO}) = \bar{Y}^2 \left[1 - \frac{1}{1 + \lambda C_y^2(1 - \rho^2)} \right] \tag{13.9}$$

Further, Grover and Kaur [22] suggested an exponential type estimator as

$$T_{GK} = [b_1\bar{y} + b_2(\bar{X} - \bar{x})]\exp\left(\frac{\bar{X} - \bar{x}}{\bar{X} + \bar{x}}\right) \tag{13.10}$$

where b_1 and b_2 are suitably chosen constants.
    For the optimum values of b_1 and b_2, the MSE of T_{GK} is derived as

$$MSE(T_{GK}) = \bar{Y}^2 \left[1 - \frac{64 + \lambda^2 C_x^2 \{16C_y^2(1 - \rho^2) + C_x^2\}}{64\{1 + \lambda C_y^2(1 - \rho^2)\}} \right] \tag{13.11}$$

Ekpenyong and Enang [3] proposed two exponential-ratio type estimators as

$$T_{PR1} = \alpha_1 \bar{y} + \alpha_2 (\bar{X} - \bar{x}) \exp\left(\frac{\bar{X} - \bar{x}}{\bar{X} + \bar{x}}\right)$$
(13.12)

and

$$T_{PR2} = \delta_1 \bar{y} + \delta_2 (\bar{X} - \bar{x}) \exp 2\left(\frac{\bar{X} - \bar{x}}{\bar{X} + \bar{x}}\right)$$
(13.13)

The MSE of T_{PR1} for optimum values of α_1 and α_2 are calculated as

$$MSE(T_{PR1}) = \bar{Y}^2 \left[1 - \frac{4 + \lambda C_x^2 (2K_{yx} - 1) + \lambda C_x^2 (1 + \lambda C_y^2)}{4(1 + \lambda C_y^2) - \lambda C_x^2 (2K_{yx} - 1)^2}\right]$$
(13.14)

Also, the MSE of T_{PR2} for optimum values of δ_1 and δ_2 is given as

$$MSE(T_{PR2}) = \bar{Y}^2 \left[1 - \frac{1 + 2\lambda C_x^2 (K_{yx} - 1) + \lambda C_x^2 (1 + \lambda C_y^2)}{1 + \lambda C_y^2 - \lambda C_x^2 (K_{yx} - 1)^2}\right]$$
(13.15)

13.3 SUGGESTED ESTIMATOR

To get a better and efficient estimate of population mean $\bar{Y}$ when $\bar{X}$ is known, we suggest a difference-cum-exponential type estimator as

$$T_s = k\bar{y} + t(\bar{X} - \bar{x})\left[1 - \exp\delta\left(\frac{\bar{X} - \bar{x}}{\bar{X} + \bar{x}}\right)\right]^{-1}$$
(13.16)

where k, t are suitably chosen constants and $\delta(\neq 0)$ is either constant or any parametric value such as β_{yx}, C_x, ρ, etc.

To derive the bias and MSE of T_s, the error terms are defined as

$$\varepsilon_1 = \frac{\bar{y} - \bar{Y}}{\bar{Y}}, \varepsilon_2 = \frac{\bar{x} - \bar{X}}{\bar{X}}$$

with $E(\varepsilon_1) = 0$, $E(\varepsilon_2) = 0$ and $E(\varepsilon_1^2) = \lambda C_y^2$, $E(\varepsilon_2^2) = \lambda C_x^2$, $E(\varepsilon_1 \varepsilon_2) = \lambda C_{xy}$.

Express T_s in terms of errors, Eq. (13.16) becomes

$$T_s = k\bar{Y}(1 + \varepsilon_1) - t\bar{X}\varepsilon_2 \left[1 - \exp\delta\left(\frac{-\varepsilon_2}{2 + \varepsilon_2}\right)\right]^{-1}$$

$$T_s = k\bar{Y}(1 + \varepsilon_1) - t\bar{X}\varepsilon_2 \left[1 - \exp\left\{-\frac{\delta\varepsilon_2}{2}\left(1 + \frac{\varepsilon_2}{2}\right)^{-1}\right\}\right]^{-1}$$

Assuming $|\varepsilon_i| < 1$, $i = 1, 2$, expand expression in terms of ε's and ignoring terms having ε's degree greater than two to get first order approximation, we have

$$T_s = k\bar{Y}(1+\varepsilon_1) - \frac{2t\bar{X}}{\delta}\left[1 - \left(1+\frac{\delta}{2}\right)\frac{\varepsilon_2}{2} + (1+\delta)\frac{\varepsilon_2^2}{4}\right]^{-1}$$

On simplifying, we get

$$T_s - \bar{Y} = \left[\bar{Y}(k-1) - \frac{2t\bar{X}}{\delta}\right] + k\bar{Y}\varepsilon_1 - t\bar{X}\left(\frac{1}{\delta}+\frac{1}{2}\right)\varepsilon_2 - \frac{t\bar{X}\delta}{8}\varepsilon_2^2 \qquad (13.17)$$

To get the bias of T_s to the first order of approximation, take expectation on both sides of Eq. (13.17), we have

$$Bias(T_s) = \left[\bar{Y}(k-1) - \frac{2t\bar{X}}{\delta}\right] - \frac{t\bar{X}\delta}{8}\lambda C_x^2 \qquad (13.18)$$

By squaring Eq. (13.17) on both sides and terminating the terms having ε's degree more than two, we have

$$(T_s - \bar{Y})^2 = \left[\bar{Y}(k-1) - \frac{2t\bar{X}}{\delta}\right]^2 + k^2\bar{Y}^2\varepsilon_1^2 - kt\bar{X}\bar{Y}\left(1+\frac{2}{\delta}\right)\varepsilon_1\varepsilon_2$$

$$+ \left[\left(\frac{3}{4}+\frac{1}{\delta}+\frac{1}{\delta^2}\right)t^2 - \frac{\delta}{4}t(k-1)R\right]\bar{X}^2\varepsilon_2^2 \qquad (13.19)$$

To get the MSE of T_s to the first order of approximation, take expectation on both sides of Eq. (13.19), we have

$$MSE(T_s) = \left[\bar{Y}(k-1) - \frac{2t\bar{X}}{\delta}\right]^2 + k^2\bar{Y}^2\lambda C_y^2 - kt\bar{X}\bar{Y}\left(1+\frac{2}{\delta}\right)\lambda C_{xy}$$

$$+ \left[\left(\frac{3}{4}+\frac{1}{\delta}+\frac{1}{\delta^2}\right)t^2 - \frac{\delta}{4}t(k-1)R\right]\bar{X}^2\lambda C_x^2 \qquad (13.20)$$

or

$$MSE(T_s) = \bar{Y}^2 \left[\left\{ (k-1) - \frac{2t}{\delta R} \right\}^2 + \lambda k^2 C_y^2 - \lambda kt C_{xy} \frac{1}{R} \left(1 + \frac{2}{\delta} \right) \right.$$

$$\left. + \left\{ \left(\frac{3}{4} + \frac{1}{\delta} + \frac{1}{\delta^2} \right) t^2 - \frac{\delta}{4} t(k-1)R \right\} \frac{\lambda C_x^2}{R^2} \right]$$

$$MSE(T_s) = \bar{Y}^2 \left[1 - 2k + k^2(1 + \lambda C_y^2) + t \left(\frac{4}{\delta R} + \frac{\lambda \delta C_x^2}{4R} \right) \right.$$

$$+ t^2 \left\{ \frac{4}{\delta^2 R^2} + \left(\frac{3}{4} + \frac{1}{\delta} + \frac{1}{\delta^2} \right) \frac{\lambda C_x^2}{R^2} \right\}$$

$$\left. - kt \left\{ \frac{4}{\delta R} + \frac{\lambda \delta C_x^2}{4R} + \lambda C_{xy} \frac{1}{R} \left(1 + \frac{2}{\delta} \right) \right\} \right] \tag{13.21}$$

or

$$MSE(T_s) = \bar{Y}^2 [1 - 2k + k^2 \xi_1 + t \xi_2 + t^2 \xi_3 - kt \xi_4] \tag{13.22}$$

where,

$$\xi_1 = 1 + \lambda C_y^2, \xi_2 = \frac{4}{\delta R} + \frac{\lambda \delta C_x^2}{4R}, \xi_3 = \frac{4}{\delta^2 R^2} + \left(\frac{3}{4} + \frac{1}{\delta} + \frac{1}{\delta^2} \right) \frac{\lambda C_x^2}{R^2},$$

$$\xi_4 = \frac{4}{\delta R} + \frac{\lambda \delta C_x^2}{4R} + \lambda C_{xy} \frac{1}{R} \left(1 + \frac{2}{\delta} \right)$$

To minimize $MSE(T_s)$, we have to differentiate $MSE(T_s)$ in Eq. (13.22) partially with respect to k and t, and equating to zero, we get

$$2\xi_1 k - \xi_4 t = 2 \tag{13.23}$$

and

$$-\xi_4 k + 2\xi_3 t = -\xi_2 \tag{13.24}$$

On solving simultaneous Eqs. (13.23) and (13.24) for k and t, the optimal values of k and t are obtained as

$$k_0 = \frac{4\xi_3 - \xi_2 \xi_4}{4\xi_1 \xi_3 - \xi_4^2}, \quad t_0 = \frac{2\xi_4 - 2\xi_1 \xi_2}{4\xi_1 \xi_3 - \xi_4^2} \tag{13.25}$$

The minimum $MSE(T_s)$ can be obtained by using the optimum values of k and t from Eq. (13.25) in Eq. (13.22), as

$$MSE_{min}(T_s) = \bar{Y}^2 \left[1 - \frac{4\xi_3 + \xi_1\xi_2^2 - 2\xi_2\xi_4}{4\xi_1\xi_3 - \xi_4^2} \right] \tag{13.26}$$

13.4 SPECIAL CASES

In this section, we do consider the different estimators with minimum of $MSE(T_s)$ by assigning the different values to δ to show the variability of the proposed estimator.

13.4.1 FOR $\delta = 1$

For $\delta = 1$, the proposed estimator T_s becomes

$$T_{s(1)} = k\bar{y} + t(\bar{X} - \bar{x}) \left[1 - \exp\left(\frac{\bar{X} - \bar{x}}{\bar{X} + \bar{x}} \right) \right]^{-1}$$

with

$$MSE_{min}(T_{s(1)}) = \bar{Y}^2 \left[1 - \frac{4\xi_{31} + \xi_{11}\xi_{21}^2 - 2\xi_{21}\xi_{41}}{4\xi_{11}\xi_{31} - \xi_{41}^2} \right] \tag{13.27}$$

where,

$$\xi_{11} = 1 + \lambda C_y^2, \xi_{21} = \frac{4}{R} + \frac{\lambda C_x^2}{4R}, \xi_{31} = \frac{4}{R^2} + \frac{11}{4}\frac{\lambda C_x^2}{R^2}, \xi_{41} = \frac{4}{R} + \frac{\lambda C_x^2}{4R} + \frac{3}{R}\lambda C_{xy}$$

13.4.2 FOR $\delta = C_x$

For $\delta = C_x$, the estimator T_s becomes

$$T_{s(2)} = k\bar{y} + t(\bar{X} - \bar{x}) \left[1 - \exp C_x \left(\frac{\bar{X} - \bar{x}}{\bar{X} + \bar{x}} \right) \right]^{-1}$$

The optimum MSE of $T_{s(2)}$ can be obtained as

$$MSE_{min}(T_{s(2)}) = \bar{Y}^2 \left[1 - \frac{4\xi_{32} + \xi_{12}\xi_{22}^2 - 2\xi_{22}\xi_{42}}{4\xi_{12}\xi_{32} - \xi_{42}^2} \right] \tag{13.28}$$

where,

$$\xi_{12} = 1 + \lambda C_y^2, \quad \xi_{22} = \frac{4}{C_x R} + \frac{\lambda C_x^3}{4R}, \quad \xi_{32} = \frac{4}{C_x^2 R^2} + \left(\frac{3C_x^2}{4} + C_x + 1 \right) \frac{\lambda}{R^2},$$

$$\xi_{42} = \frac{4}{C_x R} + \frac{\lambda C_x^3}{4R} + \lambda C_{xy} \frac{1}{R} \left(1 + \frac{2}{C_x} \right)$$

13.4.3   For $\delta = \beta_{yx}$

For $\delta = \beta_{yx}$, the estimator T_s becomes

$$T_{s(3)} = k\bar{y} + t(\bar{X} - \bar{x})\left[1 - \exp\beta_{yx}\left(\frac{\bar{X} - \bar{x}}{\bar{X} + \bar{x}}\right)\right]^{-1}$$

The minimum *MSE* of $T_{s(3)}$ is

$$MSE_{min}(T_{s(3)}) = \bar{Y}^2\left[1 - \frac{4\xi_{33} + \xi_{13}\xi_{23}^2 - 2\xi_{23}\xi_{43}}{4\xi_{13}\xi_{33} - \xi_{43}^2}\right] \tag{13.29}$$

where,

$$\xi_{13} = 1 + \lambda C_y^2, \quad \xi_{23} = \frac{4}{\beta_{yx}R} + \frac{\lambda\beta_{yx}C_x^2}{4R}, \quad \xi_{33} = \frac{4}{\beta_{yx}^2 R^2} + \left(\frac{3}{4} + \frac{1}{\beta_{yx}} + \frac{1}{\beta_{yx}^2}\right)\frac{\lambda C_x^2}{R^2},$$

$$\xi_{43} = \frac{4}{\beta_{yx}R} + \frac{\lambda\beta_{yx}C_x^2}{4R} + \lambda C_{xy}\frac{1}{R}\left(1 + \frac{2}{\beta_{yx}}\right)$$

13.4.4   For $\delta = \rho$

For $\delta = \rho$, the estimator T_s becomes

$$T_{s(4)} = k\bar{y} + t(\bar{X} - \bar{x})\left[1 - \exp\rho\left(\frac{\bar{X} - \bar{x}}{\bar{X} + \bar{x}}\right)\right]^{-1}$$

The minimum *MSE* of $T_{s(4)}$

$$MSE_{min}(T_{s(4)}) = \bar{Y}^2\left[1 - \frac{4\xi_{34} + \xi_{14}\xi_{24}^2 - 2\xi_{24}\xi_{44}}{4\xi_{14}\xi_{34} - \xi_{44}^2}\right] \tag{13.30}$$

where,

$$\xi_{14} = 1 + \lambda C_y^2, \quad \xi_{24} = \frac{4}{\rho R} + \frac{\lambda\rho C_x^2}{4R}, \quad \xi_{34} = \frac{4}{\rho^2 R^2} + \left(\frac{3}{4} + \frac{1}{\rho} + \frac{1}{\rho^2}\right)\frac{\lambda C_x^2}{R^2},$$

$$\xi_{44} = \frac{4}{\rho R} + \frac{\lambda\rho C_x^2}{4R} + \lambda C_{xy}\frac{1}{R}\left(1 + \frac{2}{\rho}\right)$$

13.5   EFFICIENCY COMPARISON

The proposed estimator T_s will be more efficient than other estimators if minimum $MSE(T_s)$ is less than MSEs of the other estimators.

For efficiency comparison, first we express the MSEs of the estimators $\bar{y}$, T_R, T_{RE}, $T_{\exp R}$, T_{RAO}, T_{GK}, T_{PR1} and T_{PR2} in terms of ξ's, we have

- $var(\bar{y}) = \bar{Y}^2(\xi_1 - 1)$

- $MSE(T_R) = \bar{Y}^2(\xi_1 + \xi_5 - 2K_{yx}\xi_5 - 1)$

- $MSE(T_{RE}) = \bar{Y}^2(\xi_1 - 1)(1 - \rho^2)$

- $MSE(T_{\exp R}) = \bar{Y}^2\left[\xi_1 - 1 + \frac{1}{4}\xi_5(1 - 4K_{yx})\right]$

- $MSE(T_{RAO}) = \bar{Y}^2\left[1 - \frac{1}{\rho^2 + \xi_1(1 - \rho^2)}\right]$

- $MSE(T_{GK}) = \bar{Y}^2\left[1 - \frac{64 + \xi_5\{16(\xi_1 - 1)(1 - \rho^2) + \xi_5\}}{64\{\rho^2 + \xi_1(1 - \rho^2)\}}\right]$

- $MSE(T_{PR1}) = \bar{Y}^2\left[1 - \frac{4 + \xi_5(2K_{yx} - 1) + \xi_1\xi_5}{4\xi_1 - \xi_5(2K_{yx} - 1)^2}\right]$

- $MSE(T_{PR2}) = \bar{Y}^2\left[1 - \frac{1 + \xi_5(K_{yx} - 1) + \xi_1\xi_5}{\xi_1 - \xi_5(K_{yx} - 1)^2}\right]$

- $MSE_{min}(T_s) = \bar{Y}^2\left[1 - \frac{4\xi_3 + \xi_1\xi_2^2 - 2\xi_2\xi_4}{4\xi_1\xi_3 - \xi_4^2}\right]$

where,

$$\xi_1 = 1 + \lambda C_y^2, \quad \xi_2 = \frac{4}{\delta R} + \frac{\lambda \delta C_x^2}{4R}, \quad \xi_3 = \frac{4}{\delta^2 R^2} + \left(\frac{3}{4} + \frac{1}{\delta} + \frac{1}{\delta^2}\right)\frac{\lambda C_x^2}{R^2},$$

$$\xi_4 = \frac{4}{\delta R} + \frac{\lambda \delta C_x^2}{4R} + \lambda C_{xy}\frac{1}{R}\left(1 + \frac{2}{\delta}\right), \quad \xi_5 = \lambda C_x^2.$$

The estimator T_s will be more efficient than estimators $\bar{y}$, T_R, T_{RE}, $T_{\exp R}$, T_{RAO}, T_{GK}, T_{PR1} and T_{PR2} whenever the following conditions satisfied,

$$MSE_{min}(T_s) - var(\bar{y}) \leq 0$$

or

$$\xi_1 + \frac{4\xi_3 + \xi_1\xi_2^2 - 2\xi_2\xi_4}{4\xi_1\xi_3 - \xi_4^2} \geq 2 \tag{13.31}$$

$$MSE_{min}(T_s) - MSE(T_R) \leq 0$$

or

$$\xi_1 + \xi_5 - 2K_{yx}\xi_5 + \frac{4\xi_3 + \xi_1\xi_2^2 - 2\xi_2\xi_4}{4\xi_1\xi_3 - \xi_4^2} \geq 2 \tag{13.32}$$

$$MSE_{min}(T_s) - MSE(T_{RE}) \leq 0$$

or

$$(\xi_1 - 1)(1 - \rho^2) + \frac{4\xi_3 + \xi_1\xi_2^2 - 2\xi_2\xi_4}{4\xi_1\xi_3 - \xi_4^2} \geq 1 \tag{13.33}$$

$$MSE_{min}(T_s) - MSE(T_{\exp R}) \leq 0$$

or

$$\xi_1 + \frac{1}{4}\xi_5(1 - 4K_{yx}) + \frac{4\xi_3 + \xi_1\xi_2^2 - 2\xi_2\xi_4}{4\xi_1\xi_3 - \xi_4^2} \geq 2 \tag{13.34}$$

$$MSE_{min}(T_s) - MSE(T_{RAO}) \leq 0$$

or

$$\frac{4\xi_3 + \xi_1\xi_2^2 - 2\xi_2\xi_4}{4\xi_1\xi_3 - \xi_4^2} \geq \frac{1}{\rho^2 + \xi_1(1 - \rho^2)} \tag{13.35}$$

$$MSE_{min}(T_s) - MSE(T_{GK}) \leq 0$$

or

$$\frac{4\xi_3 + \xi_1\xi_2^2 - 2\xi_2\xi_4}{4\xi_1\xi_3 - \xi_4^2} \geq \frac{64 + \xi_5\{16(\xi_1 - 1)(1 - \rho^2) + \xi_5\}}{64\{\rho^2 + \xi_1(1 - \rho^2)\}} \tag{13.36}$$

$$MSE_{min}(T_s) - MSE(T_{PR1}) \leq 0$$

or

$$\frac{4\xi_3 + \xi_1\xi_2^2 - 2\xi_2\xi_4}{4\xi_1\xi_3 - \xi_4^2} \geq \frac{4 + \xi_5(2K_{yx} - 1) + \xi_1\xi_5}{4\xi_1 - \xi_5(2K_{yx} - 1)^2} \tag{13.37}$$

$$MSE_{min}(T_s) - MSE(T_{PR2}) \leq 0$$

or

$$\frac{4\xi_3 + \xi_1\xi_2^2 - 2\xi_2\xi_4}{4\xi_1\xi_3 - \xi_4^2} \geq \frac{1 + \xi_5(K_{yx} - 1) + \xi_1\xi_5}{\xi_1 - \xi_5(K_{yx} - 1)^2} \tag{13.38}$$

13.6 EMPIRICAL STUDY

In this section, we have considered some real-life populations to show the efficiency of the proposed estimator over existing estimators. The description of the seven different population parameters are stated in Table 13.1.

PRE of estimators are calculated by

$$PRE(.,\bar{y}) = \frac{var(\bar{y})}{MSE(.)} \times 100 \tag{13.39}$$

It is observed from Table 13.2 that the proposed estimator T_s for different values of δ *i.e.* $\delta = 1, C_x, \beta_{yx}$ and ρ performs efficiently than the usual unbiased estimator $\bar{y}$, ratio estimator T_R, linear regression estimator T_{RE}, ratio type exponential estimator T_{expR}, Rao's difference type estimator T_{RAO}, Grover and Kaur's estimator T_{GK}, Ekpenyong and Enang's estimators T_{PR1} and T_{PR2}, respectively, for different populations. It is worth to mention that the proposed estimator $T_{s(3)}$ is more efficient than the other considered estimator(s) for Population 1 and 3. The estimator $T_{s(2)}$ is efficient for Populations 2, 4, 5, 6 and 7. Also, it is worth to mention that correlation coefficients in all considered cases are different with least in Population-6 as $\rho = 0.2522$ to highest in Population-7 as $\rho = 0.9410$ still the results are good in all cases. This shows that the proposed estimator performs better irrespective of the correlation coefficient is mild, moderate or high.

TABLE 13.1
Population Parameters

Population Source	Study Variable y	Auxiliary Variable x	N	n	$\bar{Y}$	$\bar{X}$	C_y	C_x	ρ
1. Steel and Torrie [23]	Logarithm of leaf burn in seconds	Chlorine percentage	30	6	0.6860	0.8077	0.7001	0.7493	0.4996
2. Maddala [24]	Per capita consumption of veal	Price of veal per pound	16	4	7.6375	75.4343	0.2278	0.0986	0.6823
3. Kadilar and Cingi [25]	Apple production amount	Number of apple trees	204	50	966	26441	2.4739	1.7171	0.71
4. Srivstava, Srivastava and Khare [26]	Weight of children	Mid-arm circumference of children	55	30	17.08	16.92	0.12688	0.07	0.54
5. Cochran [27]	Food cost	Income of the family	10	4	101.1	58.8	0.1449	0.1281	0.6515
6. Cochran [27]	Food cost	Income of the family	33	16	27.49	72.545	0.3685	0.1458	0.2522
7. Murthy [28]	Fixed capital	Output of factory	80	20	11.264	51.826	0.750	0.354	0.9410

TABLE 13.2

Comparison of Estimators Through PRE with Respect to $\bar{y}$

	PRE(. , $\bar{y}$)						
Estimators	Source-1	Source-2	Source-3	Source-4	Source-5	Source-6	Source-7
$\bar{y}$	100	100	100	100	100	100	100
T_R	92.10	167.59	201.55	141.13	158.82	104.49	298.97
T_{RE}	133.26	187.10	201.65	141.16	173.74	106.79	873.21
T_{expR}	133.04	133.06	159.32	128.50	161.43	106.45	163.52
T_{RAO}	139.79	188.07	210.89	141.18	174.06	107.22	875.32
T_{GK}	142.72	188.16	213.41	141.19	174.17	107.24	876.48
T_{PR1}	149.47	158.72	176.87	128.03	150.25	106.05	361.28
T_{PR2}	154.55	189.45	242.00	141.21	175.01	107.32	925.56
$T_{s(1)}$	181.95	1,177.37	407.76	597.24	227.90	1,138.57	3,204.70
$T_{s(2)}$	205.89	**2,691.97**	256.52	**1,338.97**	**478.13**	**2,223.96**	**6,277.01**
$T_{s(3)}$	**256.57**	2,521.67	**1,014.39**	602.37	193.86	2,036.63	5,744.44
$T_{s(4)}$	239.09	1,529.56	515.67	856.18	293.71	2,017.26	3,393.64

13.7   SIMULATION

Here we execute a simulation using R software to show the performance of proposed estimators over others by calculating percent relative efficiency of all estimators concerning $\bar{y}$ as defined in Eq. (13.39). For this, we have generated four hypothetical populations of size $N = 5000$ and sample size $n = 200$. Consider a linear relationship between study variable y and auxiliary variable x with slop 0.5 and drift 1. We also uses an error factor in the model $Y = 1 + 0.5X$ by random ε, which is independent of X, follows $N(0,1)$. We have considered four cases, in the first three, the auxiliary variable X follows normal distribution as $N(2,0.5)$, $N(3,1)$ and $N(4,1.5)$. In the latter case, X is taken from Uniform distribution $U(0,1)$. To make results more accurate, we have replicated it 10,000 times.

From Table 13.3, it is envisaged that the proposed class of estimators $T_{s(1)}$, $T_{s(2)}$, $T_{s(3)}$ and $T_{s(4)}$ performs efficiently in all the cases, which also supports the result obtained in numerical study in section 6.

13.8   CONCLUSION

In this paper, we propose a difference-cum-exponential type estimator for the estimation of population mean $\bar{Y}$ of the study variable y when auxiliary information is available. The bias and MSE formulae of the proposed estimator are obtained and compared with that of the usual unbiased estimator $\bar{y}$, classical ratio estimator T_R, regression estimator T_{RE}, exponential estimator T_{expR}, difference estimator T_{RAO}, Grover and Kaur's estimator T_{GK}, Ekpenyong and Enang's estimator T_{PR1} and T_{PR2}. We have examined the performance of the proposed estimator by considering seven different real-life populations and also by a simulation study. It is interesting

TABLE 13.3

PRE of Estimators with Respect to $\bar{y}$

	PRE(. , $\bar{y}$)			
Estimators	Case-1	Case-2	Case-3	Case-4
$\bar{y}$	100	100	100	100
T_R	98.3515	109.3995	132.2530	76.2459
T_{RE}	105.3458	122.7273	152.1649	101.9720
T_{expR}	105.3127	122.2299	148.0307	97.2554
T_{RAO}	105.4733	122.8233	152.2483	102.2855
T_{GK}	105.4813	122.8404	152.2759	102.3278
T_{PR1}	105.9562	119.7086	139.3933	111.1496
T_{PR2}	105.5044	122.9139	152.4357	102.3529
$T_{s(1)}$	770.9312	326.5865	222.2267	404.7270
$T_{s(2)}$	1,346.4636	529.0413	351.6292	522.1004
$T_{s(3)}$	1,126.0685	468.4702	319.4904	555.5838
$T_{s(4)}$	1,376.1330	487.9484	295.5479	726.9252

to remark that the proposed class of estimators performs efficiently than the other considered estimators. Thus, our conclusion is to recommend the proposed estimator for future study.

REFERENCES

1. Kadilar, C. and Cingi, H. 2004. Ratio estimators in simple random sampling. *Applied Mathematics and Computation.* 151(3):893–902.
2. Upadhyaya, L. N. and Singh, H. P. 1999. Use of transformed auxiliary variable in estimating the finite population mean. *Biometrical Journal.* 41(5):627–636.
3. Ekpenyong, E. J. and Enang, E. I. 2015. Efficient exponential estimator for estimating the population mean in simple random sampling. *Hacettepe Journal of Mathematics and Statistics.* 40(3):689–705.
4. Kadilar, C. and Cingi, H. 2006. Improvement in estimating the population mean in simple random sampling. *Applied Mathematics Letter.* 19(1):75–79.
5. Singh, H. P. and Solanki, R. S. 2012. An alternative procedure for estimating the population mean in simple random sampling. *Pakistan Journal of Statistics and Operations Research.* VIII(2):213–232.
6. Ray, S. K. and Sahai, A. 1980. Efficient families of ratio and product type estimators. *Biometrika.* 67(1):211–215.
7. Pandey, B. N. and Dubey, V. 1988. Modified product estimator using coefficient of variation of auxiliary variate. *Assam Statistical Review.* 2(2):64–66.
8. Reddy, V. N. 1973. On ratio and product method of estimation. *Sankhya. Series B.* 35:307–316.
9. Singh, H. P. and Espejo, M. R. 2003. On linear regression and ratio-product estimation of a finite population mean. *The Statistician.* 52(1):59–67.
10. Singh, H. P. and Tailor, R. 2005, Estimation of finite population mean with known coefficient of variation of an auxiliary character. *Statistica.* 65(3):301–313.

11. Singh, H. P. and Agnihotri, N. A. 2008. A general procedure of estimating population mean using auxiliary information in sample survey. *Statistics in Transition-new series.* 9, 71–87.
12. Tailor, R. and Sharma, B. A. 2009. Modified ratio-cum-product estimator of finite population mean using known coefficient of variation and coefficient of kurtosis. *Statistics in Transition-New Series.* 10, 15–24.
13. Singh, H. P. and Solanki, R. S. 2011. Generalized ratio product methods of estimation in survey sampling. *Pakistan Journal of Statistics and Operation Research.* 7(2):245–264.
14. Solanki, R. S., Singh, H. P. and Pal, S. K. 2013. Improved estimation of finite population mean in sample survey. *Journal of Advance Computing.* 1, 70–78.
15. Kumar, S. 2013. Estimation in finite population surveys: Theory and applications. *Journal of Modern Applied Statistical Method.* 12(1)120–127.
16. Pal, S. K., Singh, H. P., Kumar, S. and Chatterjee, K. A. 2017. A family of efficient estimators of the finite population mean in simple random sampling. *Journal of Statistical Computation and Simulation.* 88(5):920–934.
17. Bahl, S., Tuteja, R. K. 1991. Ratio and product type exponential estimators. *Journal of Information and Optimization Sciences.* 12, 159–164.
18. Gupta, S. and Shabbir, J. 2008. On the improvement in estimating the population mean in simple random sampling. *Journal of Applied Sciences.* 35(5):559–566.
19. Solanki, R. S., Singh, H. P. and Rathour, A. 2012. An alternative estimator for estimating finite population mean using auxiliary information in sample surveys. *ISRN Probability and Statistics.* 1–14.
20. Yadav, S. K. and Kadilar, C. 2013. Efficient familiy of exponential estiamtors for the population mean. *Hacettepe Journal of Mathematics and Statistics.* 42(6):671–677.
21. Rao, T. J. 1991. On certain methods of improving ratio and regression estimators. *Communications in Statistics-Theory and Methods.* 20(10):3325–3340.
22. Grover, L. K. and Kaur, P. 2011. An improved estimator of the finite population mean in simple random sampling. *Model Assisted Statistics and Applications.* 6(1):47–55.
23. Steel, R., Torrie, J. and Dickey D. 1960. Principles and procedures of statistics. Detroit, MI: McGraw-Hill Companies.
24. Maddala, G. S. 1977. Econometrics, economics handbook series. New York, NY: McGraw-Hills Publication Company.
25. Kadilar, C. and Cingi, H. 2005. A new estimator using two auxiliary variables. *Applied Mathematics and Computation.* 151, 893–902.
26. Srivastava, R. S., Srivastava, S. P. and Khare, B. B. 1989. Chain ratio type estimator for ratio of two population means using auxiliary characters. *Communication in Statistic Theory and Methods.* 18(10):3917–3926.
27. Cochran, W. G. 1977. Sampling techniques. 3rd ed. New York, NY: Wiley.
28. Murthy, M. N. 1967. Sampling theory and methods. Calcutta, India: Statistical Publishing Society.

14 A Study of After-Effects of Kerala Floods Using VIIRS-OLS Nighttime Light Data

Pranjal Dave, Sumanta Pasari
Birla Institute of Technology and Science, Pilani,
Rajasthan, India

CONTENTS

14.1 Introduction 195
14.2 Dataset and Study Area 196
14.3 Methodology 197
    14.3.1 Preprocessing.. 197
    14.3.2 Processing... 197
    14.3.3 Postprocessing 198
14.4 Results and Discussions... 199
References... 201

14.1 INTRODUCTION

Remote sensing is a widely used technique for the study of the Earth's surface features. Nighttime light imagery forms a useful data set in remote sensing. It is an essential dataset for the study of human activities on Earth and their effects. Such datasets are also found to be of great importance in the study of the atmosphere and other natural processes. A few common examples are the detection and monitoring of city lights, fires, dust storms, volcanoes, gas flares and population/economic geography (Chuvieco 2016). As the name suggests, nighttime light imagery data is collected during the night. It usually contains two types of features: self-illuminating features and moon-illuminated features. Some examples of self-illuminating features include gas flares, forest fires, human-caused disasters, volcanic lava and bioluminescence, whereas moonlight illuminating features include snow cover, sea ice and volcanic ash along with various surface features such as mountains, deserts, rivers and moon glint. Another source of illumination for the clouds can be airglow. It is the luminosity occurring due to chemical reactions in the upper atmosphere. Nighttime visible imaging was initiated by the Defense Meteorological Satellite Program – Operational Linescan System (DMSP-OLS) in the 1960s, which was the only source of visible nighttime images until the launch of Suomi National Polar-Orbiting Partnership – Visible Infrared Imaging Radiometer Suite (SNPP-VIIRS) in 2011. The VIIRS instrument collects visible and infrared imagery and global observations of the land, atmosphere, cryosphere and oceans. In Doll (2008) and Elvidge et al. (2017), a detailed explanation of the generation and usage of the Visible Infrared Imaging Radiometer Suite Day/Night Band (VIIRS-DNB) satellite data is available.

Several studies have highlighted the use of nighttime light data for efficient disaster management. For example, Gillespie et al. (2014) have utilized the nighttime light imagery from the DMSP-OLS to study the effect of the 2004 Indian Ocean mega-tsunami in Indonesia. Their findings were validated based on the extensive aftermath survey results. An empirical relationship between economic expenditure after the tsunami and the nighttime light data was established (Gillespie et al. 2014). Similarly, Zhao et al. (2018) used the National Polar-Orbiting Partnership Visible Infrared Imaging Radiometer Suite Day/Night Band (NPP-VIIRS DNB) daily data to study the effect of three major disasters, namely the earthquake, storms and floods. The percent-of-normal light (PNL) method was employed to assess the damage caused by the disaster. The NPP-VIIRS DNB data one month prior to disaster and ten days after the disaster was averaged to obtain predisaster and postdisaster values. In their study, a longer time period for predisaster was chosen so as to account for the variations of the light intensity due to clouds and other effects, whereas a shorter postdisaster period was chosen to represent a critical period after which postdisaster activity occurs. The nonparametric Mann-Whitney U-Test was employed to test the statistical significance of the results. In an another effort, Wang et al. (2018) studied the power outages in the United States as an aftermath of Hurricane Sandy in 2012 and Hurricane María in 2017 using NASA black marble product data that removes cloud-contaminated pixels and corrects for atmospheric, terrain, vegetation, snow, lunar and stray light effects on the VIIRS DNB radiances. Percent-of-normal light technique was used to assess the damage.

Motivated by the above applications of nighttime light data in disaster assessment and aftermath management, the present study aims to estimate flood-related damage in the state of Kerala using VIIRS-DNB scan data. The floods that occurred in the month of August 2018 in Kerala were caused by the unusually high rainfall during the monsoon season. They were one of the worst floods in Kerala in nearly a century, causing property damage of about ₹400 billion (US$5.8 billion) and casualty of over 483 human deaths in the state.

14.2   DATASET AND STUDY AREA

For the present study, we use the DNB sensor data of VIIRS instrument (VIIRS-DNB) provided by the Earth Observations Group (EOG) at NOAA/NCEI. Table 14.1 summarizes some key specifications of the sensor. The VIIRS-DNB data is mainly of three types: stable lights, radiance calibrated and the average digital number based on the detection frequency, radiance and digital number, respectively. These differences make the datasets usable in several applications, such as the study of urban extent, population change over time, socio-economic activity, greenhouse

TABLE 14.1
VIIRS-DNB Sensor Specification

Coverage	22 spectral bands from 412 nm to 12 μm
Nadir Resolution	400 m
Swath	3,000 km (max)
Average Data Rate	7,674,000 bps
Average Power	319 Watts

gas emissions, light pollution and disaster management. In this work, the study region consists of the state of Kerala, a coastal state located in the southwestern part of India.

14.3 METHODOLOGY

The methodology comprises three steps: preprocessing of data, processing and post-processing analysis.

14.3.1 PREPROCESSING

When a satellite records data, some phenomena such as stray lights, reflected moonlight from clouds, snow cover, topographic variations and the position of the pixels in the swath substantially affect the recorded radiance value. To account for such disturbances, lunar radiance removal, cloud clearing and edge-of-swath pixel removal are often used. To preprocess the data for the present work, we use the summed cloud cover (SCC) recorded by the satellite and the average-swath technique that considers the aggregated value of swath from 32 different zones on each side of the nadir via a scan-angle-dependent aggregation strategy.

14.3.2 PROCESSING

To enable rapid detection of the flood affected areas in the state of Kerala, we have chosen the percent of normal light (PNL) technique, which is based on the realization that the flood-affected areas will have direct impact on the urban lights. As the nighttime light data captures these lights, they are used to develop the PNL method. To detect the change, the radiance values before and after the disaster are compared by taking their ratio. If the ratio is less than 1, the area is classified as affected. The PNL technique, due to its simplicity and availability of relevant dataset, is often preferred for rapid damage assessment over other methods, such as the false color composite, Flood Proxy Maps (FPM) and estimation using economic parameters.

To implement the PNL technique, VIIRS-DNB scan data for the months of July, August and September was analyzed. The radiance values were first thresholded to ensure a correct calculation of the PNL values, as for very small radiance values, background noise can result in extremely high or extremely low values of PNL. Thus, Rad_{pre} and Rad_{post} values below 0.3 nW.cm^{-2}.sr^{-1} have been thresholded. The accrued monthly data was then averaged to generate the composite radiance values. As the floods occurred in the month of August, the data for the month of August was taken to compute postdisaster radiance values. The data for the months of July and September are averaged and taken as predisaster radiance values. Using the above values, the PNL image for the study region was generated by taking the ratio of pre-disaster radiance values and postdisaster radiance values as

$$PNL = \frac{Rad_{post}}{Rad_{pre}} \qquad (14.1)$$

14.3.3 POSTPROCESSING

After obtaining the PNL image, a simple criterion of PNL value less than 1 is employed to demarcate the affected areas. District-level sum of lights statistics is then generated using the composite radiance values. Sum of lights is defined as the sum of all the radiance values in the district. The obtained values are plotted on a histogram to observe the drop in radiance values after the disaster. The districts with the large drop in radiance values were identified as worst affected.

A flow-chart in Figure 14.1 describes the above methodology of flood damage assessment in the state of Kerala.

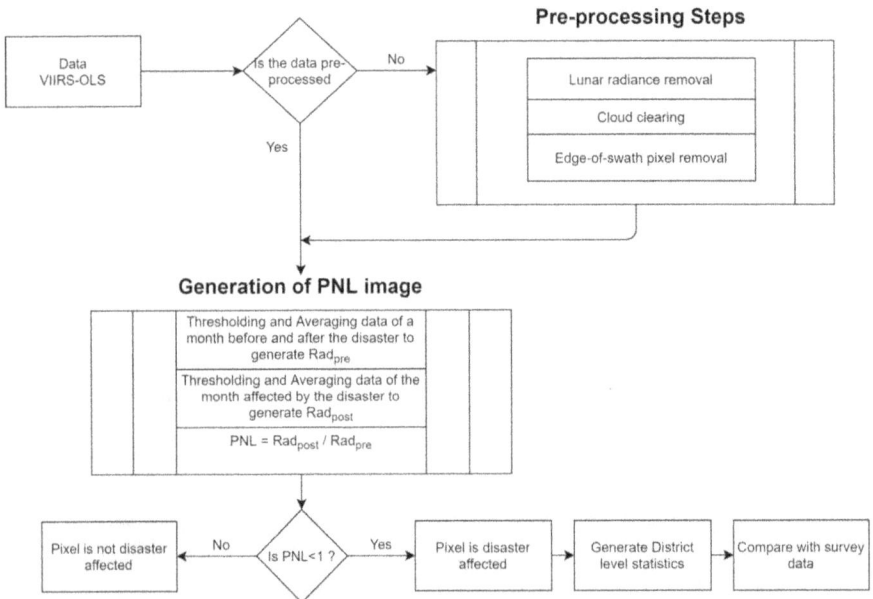

FIGURE 14.1 Flow-chart of the proposed methodology.

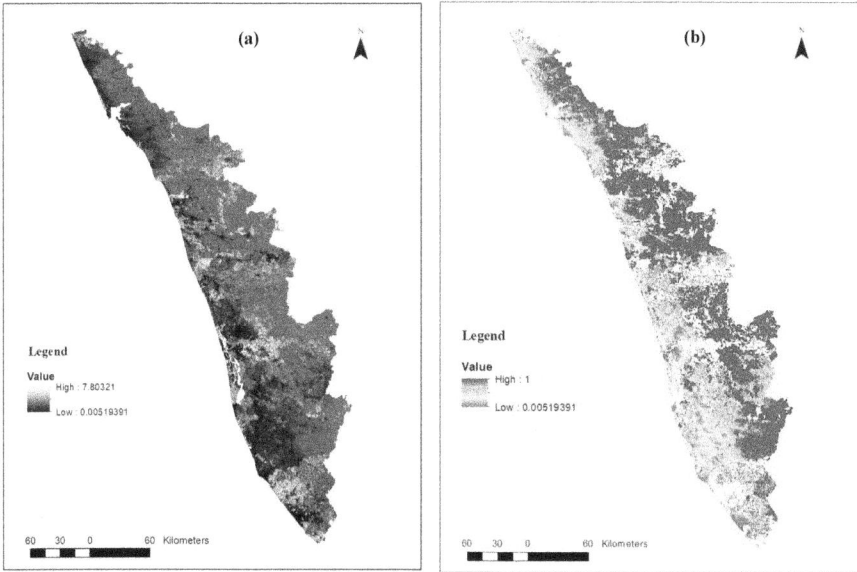

FIGURE 14.2 (a) The PNL values of Kerala state and (b) the 2018 flood-affected areas (PNL < 1).

14.4 RESULTS AND DISCUSSIONS

In this section, we present the results. The image of PNL values for the study region is shown in Figure 14.2a. Areas with low PNL values (less than 1) suggest that the region is highly affected by the floods, as the lights coming from these areas after the flood have decreased by a larger amount. The extent of damage by floods are represented in a color-coded image as shown in Figure 14.2b, whereas a summary of PNL values is provided in Table 14.2.

The predisaster and postdisaster images were used to calculate district-wise statistics for Kerala. The predisaster and postdisaster radiance values for each district were summed and represented in Figure 14.3a and Figure 14.3b, respectively. Using the radiance-sum value for each district, a graph (Figure 14.4) was generated to highlight the change in the predisaster and postdisaster radiance-sum of lights for each district.

From the graph in Figure 14.4, it is observed that the postdisaster values of the sum of lights are less than those for the predisaster values. According to ground surveys

TABLE 14.2
Summary of Results

% area affected	49.04
Number of affected pixels (out of 179,310)	8,7940
Mean PNL value	0.428
Range of PNL values	0–5.014

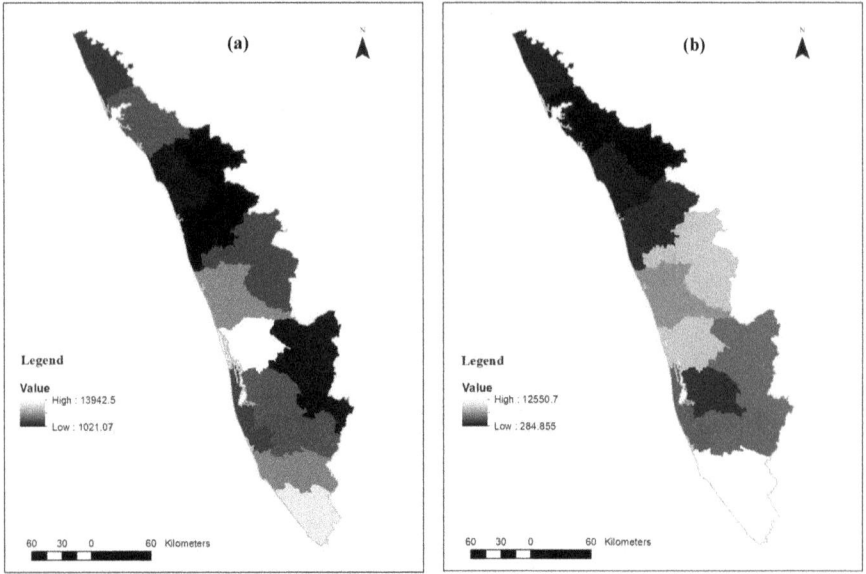

FIGURE 14.3 District-wise sum of lights in Kerala; (a) pre-disaster sum of lights and (b) post-disaster sum of lights.

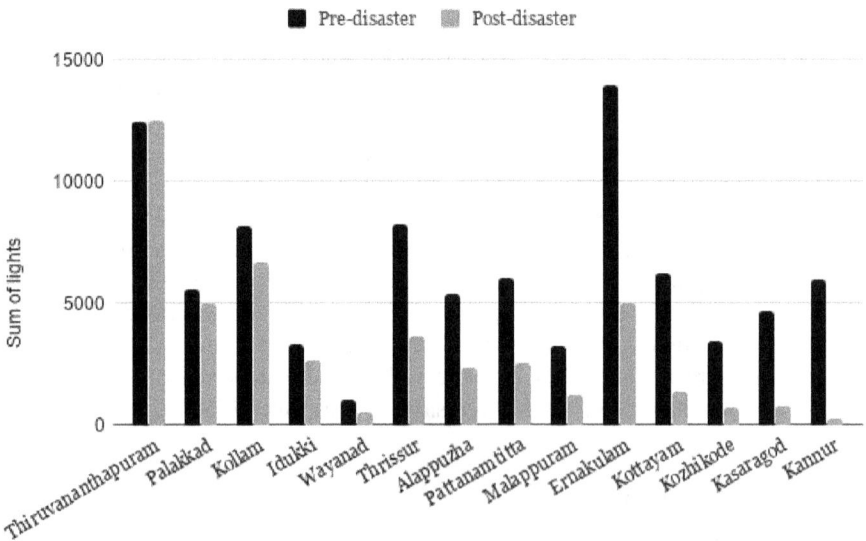

FIGURE 14.4 Bar plots for pre-disaster and post-disaster radiance-sum of lights for the districts of Kerala.

carried out by the Indian government and various agencies (Kerala Floods 2018), the worst affected areas of the state were Wayanad, Pathanamthitta, Ernakulam, Thrissur, Malappuram, Kozhikode, Kannur, Palakkad and Alappuzha. It can be observed from Figure 14.4 that there is a significant drop in the radiance values for these districts except for Wayanad. Due to insufficient radiance values even before the disaster, the effect of flood measured through the sum of lights is not well represented for Wayanad district.

In summary, in this chapter, we have used the concept of percent of normal light to identify disaster-affected areas in the state of Kerala through the nighttime light image. We further calculated the sum of lights for each district and plotted the results. We have not involved any economic parameters in the analysis as the analysis is carried out for a short duration of time around the disaster. An economic parameter will not be a suitable indicator for such a duration. A significant drop was observed in the values of the postdisaster sum of lights. The districts with the most drop were identified as severely affected. These results were compared with the ground reports and were found to be valid. Thus, the proposed analysis is found to be useful for rapid and efficient flood damage assessment in an area.

Processing system: The system used to generate the results has an Intel i7 8th gen CPU unit along with a NVIDIA GeForce GTX 1050 Ti (4 GB) GPU. The system has a DDR4 RAM of 16 gigabytes. We use ArcGIS (v10.2.2) suite of geospatial processing software for data processing.

REFERENCES

Chuvieco, E. 2016. *Fundamentals of Satellite Remote Sensing: An Environmental Approach.* Boca Raton, FL: CRC Press, Taylor & Francis Group.

Doll, C.N.H. 2008. CIESIN thematic guide to night-time light remote sensing and its applications. New York, NY: Center for International Earth Science Information Network of Columbia University. Retrieved from http://sedac.ciesin.columbia.edu/tg/ (Accessed on 27 November 2019).

Elvidge, C.D., Baugh, K., Zhizhin, M., Hsu, F.-C. and Ghosh, T. 2017. Supporting international efforts for detecting illegal fishing and GAS flaring using VIIRS. *2017 IEEE International Geoscience and Remote Sensing Symposium (IGARSS).* Doi: 10.1109/igarss.2017.8127580.

Gillespie, T.W., Frankenberg, E., Chum, K.F. and Thomas, D. 2014. Night-time lights time series of tsunami damage, recovery, and economic metrics in Sumatra, Indonesia. *Remote Sensing Letters.* 5(3), 286–294. Doi: 10.1080/2150704x.2014.900205.

Kerala Floods. 2018. These are the worst affected areas, stay clear of them. Mirror Now News (website). https://www.timesnownews.com/mirror-now/in-focus/article/kerala-floods-2018-these-are-the-worst-affected-areas-stay-clear-of-them/271873.

Wang, Z., Román, M.O., Sun, Q., Molthan, A. L., Schultz, L. A. and Kalb, V. L. 2018. Monitoring disaster-related power outages using NASA black marble nighttime light product. *ISPRS - International Archives of the Photogrammetry, Remote Sensing and Spatial Information Sciences.* XLII-3, 1853–1856. Doi: 10.5194/isprs-archives-xlii-3-1853-2018.

Zhao, X., Yu, B., Liu, Y., Yao, S., Lian, T., Chen, L., … Wu, J. 2018. NPP-VIIRS DNB daily data in natural disaster assessment: Evidence from selected case studies. *Remote Sensing.* 10(10), 1526. Doi: 10.3390/rs10101526.

15 Indian Plate Motion Revealed by GPS Observations: Preliminary Results

Yogendra Sharma, Sumanta Pasari and Neha
Birla Institute of Technology and Science, Pilani,
Rajasthan, India

CONTENTS

15.1 INTRODUCTION

The Indian plate is one of the most active tectonic plates in the world. This plate is colliding with the Eurasian plate since 55 Ma [Yin, 2006]. Due to this persistent collision, many types of tectonic hazards (e.g., earthquakes, volcanoes, landslides) have occurred along the plate boundary as well as in the plate interior. This collision created the world's largest mountain range, the Himalayas, which has deformed many times due to several large earthquakes, such as the 1905 Kangra earthquake (Mw = 7.8), 1934 Nepal-Bihar earthquake (Mw = 8.1), 1950 Assam earthquake (Mw = 8.4), 1991 Uttarkashi earthquake (Mw = 6.8), 1999 Chamoli earthquake (Mw = 6.8), 2005 Kashmir earthquake (Mw = 7.6) and the 2015 Gorkha earthquake (Mw = 7.8), causing millions of deaths along this arc and its surrounding Indo-Gangetic plains (Figure 15.2) [Ambraseys and Douglas, 2004; Avouac et al., 2015; Bilham, 2019; Kaneda et al., 2008]. Apart from these interplate earthquakes, the Indian plate has also experienced some devastating earthquakes at its interior part, such as the 1967 Koyna earthquake (Mw = 6.6), 1993 Latur earthquake (Mw = 6.2) and the 2001 Bhuj earthquake (Mw = 7.7) (Figure 15.2) [Bilham et al., 2003]. While geodetic measurements across the Indian plate suggest that the Indian continent behaves as a stable shield, the appearance of notable events along the central part of India also intimates minor deformation (~ 3 ± 2 mm/yr) within the Indian subcontinent [Bilham et al., 1998; Gupta, 1993; Jade et al., 2004; Paul et al., 2001].

There have been several studies on Indian plate motion and its deformation [Bilham et al., 1998; Jade et al., 2004, 2017; Paul et al., 2001]. For instance, Bilham et al. (1997) suggested that ~20 mm/yr of total convergence between the Indian and Eurasian plate has been observed along the Himalaya [Bilham et al., 1997]. Similarly, the geological studies along the Main Himalayan Thrust (MHT) also suggest that about 50 percent of the convergence is absorbed by the Himalayas [Lavé and Avouac, 2000]. Lavé and Avouac (2000) have reported that the maximum shortening along the Himalayas has concentrated well north of the surface trace of MHT [Lavé and Avouac, 2000]. Using the observations from 50 GPS sites, Jade et al. (2004) concluded that the peninsular India moves as a rigid plate, while about $\sim 10 - 20$ mm/yr convergence occurs along the Himalayan arc [Jade et al., 2004]. Banerjee et al. (2008) collected GPS data across the Indian subcontinent and suggest that the whole of central India accommodates about $\sim 2 - 1$ mm/yr convergence [Banerjee et al., 2008]. Mahesh et al. (2012) have suggested that the Indian subcontinent is deforming with a shallow rate ($< 1 - 2$ mm/yr), and the whole plate interior acts like a solid plate [Mahesh et al., 2012]. Similarly, Jade et al. (2017) have also estimated the intraplate deformation rate of the Indian plate about $\sim 1 - 2$ mm/yr [Jade et al., 2017].

In the current study, we have used four years of GPS data from four continuous International GNSS Service (IGS) stations (three from the Indian plate and one is from the Eurasian plate) to estimate the present-day velocity field of the Indian plate in order to constrain the intraplate as well as the interplate crustal deformation of the Indian subcontinent.

15.2   GPS OVERVIEW

Global Positioning System (GPS) is a space-based navigation system stabilized by the US Department of Defense (DoD). GPS is composed of three main segments: the space segment, the control segment and the user segment. The space segment comprises 31 satellites placed in six different orbital planes at an inclination of 55° and elevation of 20,200 km above the Earth's surface [Hofmann et al., 2012]. Each satellite transmits the data at two different carrier frequencies of L1 = 1,575.42 MHz and L2 = 1,227.69 MHz. The L1 band carries the navigation message, which consists of the ephemerides information, predicted GPS satellite orbits, clock corrections, ionospheric noise and satellite health status [French, 1996; Van, 2009]. The control segment contains one master control station (MCS), five monitoring stations and four ground antenna. The main jobs of this segment are tracing the satellite orbit, determining clock corrections and formulation of the navigation data. The user segment includes the GPS receivers that use the received information from the satellites to calculate its position and time [Hofmann et al., 2012]. The clock reading at the satellite antenna is compared with a clock reading at the receiver antenna. This comparison provides the distance from receiver antenna to satellite (pseudorange) and the time of traveling of the signal between satellite and receiver with the multiplication of speed of light [Hofmann et al., 2012; Van, 2009]. The pseudorange can be displayed as

$$R = \rho_r^s + c\Delta\delta + d_{ion} + d_{trop} + d_{tide} + \varepsilon_p \tag{15.1}$$

Here,

$$\rho_r^s = \sqrt{\left(\left(X^s(t) - X_r\right)^2 + \left(Y^s(t) - Y_r\right)^2 + \left(Z^s(t) - Z_r\right)^2\right)} \tag{15.2}$$

is the geometric range between satellite and receiver antenna; $X^s(t)$, $Y^s(t)$, $Z^s(t)$ are the components of the geocentric position vector of the satellite at epoch t; X_r, Y_r, Z_r are the three coordinates of the observing receiver; c is the speed of light; $\Delta\delta$ is the offset between the receiver clock and satellite clock; d_{ion}, d_{trop} and d_{tide} are the ionospheric delays, tropospheric delays and loading of tide effects, respectively; and ε_p represents the effect of multipath and receiver noise [French, 1996; Grewal et al., 2007; Hofmann et al., 2012; Van, 2009].

On the other hand, the carrier phase is a measure of the phase difference between the received carrier and signal generated by the GPS receiver. Positioning accuracy from the carrier phase (ϕ) is many times better than the accuracy of code pseudoranges. The carrier phase equation can be represented as follows

$$\lambda\phi = \rho_r^s + c\Delta\delta + \lambda N + d_{ion} + d_{trop} + d_{tide} + \varepsilon_p \tag{15.3}$$

Here, N is the ambiguity related to the receiver and satellite (number of fractional phases), and λ is the carrier wavelength [French, 1996; Grewal et al., 2007; Hofmann et al., 2012; Van, 2009]. There are many sources of error that could affect the accuracy of the GPS observations, namely the ionospheric/tropospheric delays, satellite orbital errors, ocean tide loading effect, receiver and satellite clock biases and multipath noises. To reduce these errors in the estimation of GPS coordinates and relative velocity, the linear combination approach is used in the present analysis. The receiver and satellite clock biases can be reduced using the double-difference method [Hofmann et al., 2012]. To understand the double-difference method, let us assume two receivers a, b, and two satellites j, k. Two carrier phase observation equations according to Eq. (15.3) can be written as:

$$\lambda\phi_a^j = \rho_a^j + c\Delta\delta_a + \lambda N_a^j + d_{a\ ino}^j + d_{a\ trop}^j + d_{a\ tide}^j + \varepsilon_{a\ p}^j \tag{15.4}$$

$$\lambda\phi_b^j = \rho_b^j + c\Delta\delta_b + \lambda N_b^j + d_{b\ ino}^j + d_{b\ trop}^j + d_{b\ tide}^j + \varepsilon_{b\ p}^j \tag{15.5}$$

First, we perform the single difference for satellite j and receivers a and b by subtracting Eq. (15.4) from Eq. (15.5)

$$\lambda\phi_{ab}^j = \rho_{ab}^j + c\Delta\delta_{ab} + \lambda N_{ab}^j + d_{ab\ ino}^j + d_{ab\ trop}^j + d_{ab\ tide}^j + \varepsilon_{ab\ p}^j \tag{15.6}$$

Similarly, the single difference for satellite k and receivers a and b is

$$\lambda \phi_{ab}^k = \rho_{ab}^k + c\Delta\delta_{ab} + \lambda N_{ab}^k + d_{ab\ ino}^k + d_{ab\ trop}^k + d_{ab\ tide}^k + \varepsilon_{ab\ p}^k \qquad (15.7)$$

To obtain double-difference, we have subtracted these single-difference equations (Eqs. [15.6] and [15.7])

$$\phi_{ab}^{jk} = \frac{1}{\lambda}\rho_{ab}^{jk} + N_{ab}^{jk} + \frac{1}{\lambda}\left(d_{ab\ ino}^{jk} + d_{ab\ trop}^{jk} + d_{ab\ tide}^{jk} + \varepsilon_{ab\ p}^{jk}\right) \qquad (15.8)$$

The advantage of double difference is that the receiver clock biases are completely eliminated and the ionospheric and tropospheric effects are reduced to a great extent [French, 1996; Grewal et al., 2007; Hofmann et al., 2012; Van, 2009]. These corrected GPS observations are now used to calculate the position and relative velocity of the receiver.

15.3   GPS DATA PROCESSING

For the present study, we have accrued four years $(2015-2019)$ of GPS data from three IGS stations (IISC, HYDE and LCK4) from the Indian plate and one IGS station (LHAZ) from the Eurasian plate along with four additional IGS stations (CHUM, KIT3, POL2 and URUM) from Scripps Orbit and Permanent Array Center (SOPAC). GPS data is generally stored in RINEX (Receiver Independent Exchange) format. The RINEX files are further used for data processing. For high precision research work in geodesy, standard scientific GPS postprocessing software (GAMIT/GLOBK, BERNESE and GIPSY) is utilized. In the present study, we have used GAMIT/GLOBK postprocessing software to analyze the available GPS data. GAMIT/GLOBK is available on the LINUX environment [Herring et al., 2010]. This software is a GPS data processing software developed by the Massachusetts Institute of Technology (MIT) for the estimation of three-dimensional relative positions of a ground station. GAMIT uses GPS broadcast carrier phase and pseudorange observables (stored in RINEX file), also known as GPS readings, satellite ephemeris (stored in navigation file) and satellite orbit data (stored in orbit file). Through the least-squares estimation, it generates values of positions and other parameters (orbits, Earth orientation, ambiguities and atmospheric delays) [Herring et al., 2010; Leberl, 1978]. We have derived the position of GPS station from Eq. (15.2). The linearized form of the equation allows us to implement the least-squares algorithm. The simplified and linear form of Eq. (15.2) is given below

$$d = Ax + v \qquad (15.9)$$

where
$\quad d\ [n \times 1]$ = vector of observations
$\quad A\ [n \times u]$ = design matrix
$\quad x\ [u \times 1]$ = vector of unknowns (parameter)
$\quad v\ [n \times 1]$ = noise or residual vector

For further computation, let us define some additional parameters,

σ_0^2 = a priori variance

Σ = covariance matrix

$Q_d = \dfrac{1}{\sigma_0^2}\Sigma$ = the cofactor matrix of observations

$P = Q_d^{-1}$ = the weight matrix

The least-squares adjustment provides a unique solution of Eq. (15.9)

$$v^T P v = \text{minimum.}$$

This adjustment principle provides following normal equation:

$$A^T P A x = A^T P d \qquad (15.10)$$

The solution of Eq. (15.10) is

$$x = \left(A^T P A\right)^{-1} A^T P d, \qquad (15.11)$$

which can be simplified to

$$x = G^{-1} g, \qquad (15.12)$$

where

$G = A^T P A$

$g = A^T P d$

The cofactor matrix Q_x follows from $x = G^{-1} A^T P d$ by the covariance propagation law as

$$Q_x = \left(G^{-1} A^T P\right) Q_d \left(G^{-1} A^T P\right)^T \qquad (15.13)$$

and further reduces to

$$Q_x = G^{-1} = \left(A^T P A\right)^{-1} \qquad (15.14)$$

by substituting $Q_d = P^{-1}$. The daily solutions from GAMIT provide the location coordinates for each station along with the Earth orientation and satellite orbit corrections. Further, the estimated loosely constrained daily solutions have been utilized to estimate the station position and plate motion using GLOBK [Herring et al., 2010]. GLOBK suite takes results from GAMIT solution files (called h-files) and daily solution of global IGS stations processed and archived at SOPAC and merges them together with a Kalman Filter estimator to provide the GPS time series and velocity for all the GPS stations [Herring et al., 2010]. However, GLOBK assumes a linear model, which cannot correct any deficiency of initial loosely constrained solution (h-file). To further identify and remove any measurements or stations which are outliers, we have used GG-MATLAB (GAMIT/GLOBK MATLAB) toolbox [Herring et al., 2010]. Once all corrections and refinement of data are made, we filter the data through GLOBK to obtain the station velocity. Further, we discussed the time series of each station with the seasonal component and velocity estimation for all four stations.

15.4   TIME SERIES ANALYSIS

The final estimated daily positions at each site were transformed into the International Terrestrial Reference Frame 2008 (ITRF08) for further analysis [Altamimi et al., 2012]. Figure 15.1 represents the time series result in the north, east, and upward direction of each station. The discontinuities or jumps that occur in the GPS position time series are probably due to the multipath effect, antenna error or the seasonal variation. The seasonal variation is found to be significant in the vertical component of displacement vectors, whereas minor impact can be observed in the north and east components for all the stations (Figure 15.1). The modulation of seasonal variation can be the combination of surface loading related to water variations, ionospheric-tropospheric pressure, vapor loading during the winter season (Dam et al., 2001). The seasonal effect can be decomposed into annual and semiannual components. These components can be represented into the linear function of sine and cosine period terms:

$$y(t) = a + b \times t + c \times \cos\left(\frac{2\pi t}{T}\right) + d \times \sin\left(\frac{2\pi t}{T}\right) + e \times \cos\left(\frac{4\pi t}{T}\right) + f \times \sin\left(\frac{4\pi t}{T}\right)$$
$$(15.15)$$

Here, a is the intercept (constant value), b is the secular rate, c and d are the amplitude of annual (12 months) periodic perturbations (sine and cosine terms) and e and f are the amplitude of semiannual (six months) periodic disturbances (sine and cosine terms). We used the GG-MATLAB toolbox to derive the seasonal variation from the GPS time series using the MATLAB function called tsview. The amplitude of the annual seasonal effect is lying in the range of ~0.3 mm to ~1.7 mm, ~0.2 mm to 0.9 mm and ~1.3 mm to ~2.1 mm for the north, east and vertical component, respectively, for all the four stations. The amplitude of the semi-annual seasonal effect is usually lesser than the amplitude of annual seasonal effects (Blewitt and Lavallee, 2002). It has been noted that continuous observations for a longer time span (>2.5 years) reduce the influence of seasonal variation in the estimation of station velocity (Blewitt and Lavallee, 2002). The coordinate time series and velocities derived from GAMIT/GLOBK are shown in Figure 15.1 and Table 15.1, respectively, in the ITRF08 reference frame.

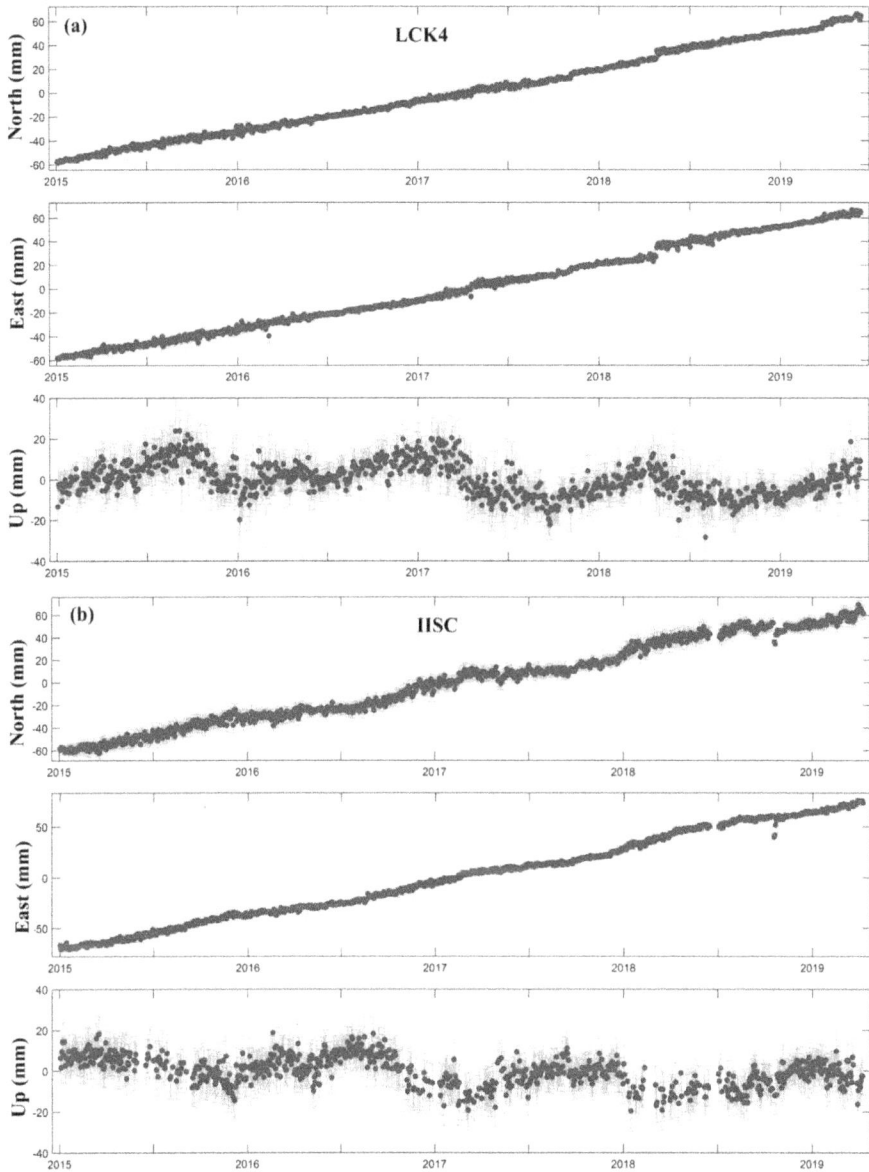

FIGURE 15.1 Time series plot of coordinates of all the continuous IGS stations (a) LCK4, (b) HYDE, (c) IISC and (d) LHAZ). The data gaps in the time series plot of all stations are may be due to signal obstructions or electricity failure. First two plots for all the stations show the linear trend along the northern and eastern direction, respectively, and the third plot for the station represents the vertical displacement factor in the data. The blue dots are daily position of each GPS station in the north, east and the upward direction along with their uncertainties (light black bar).

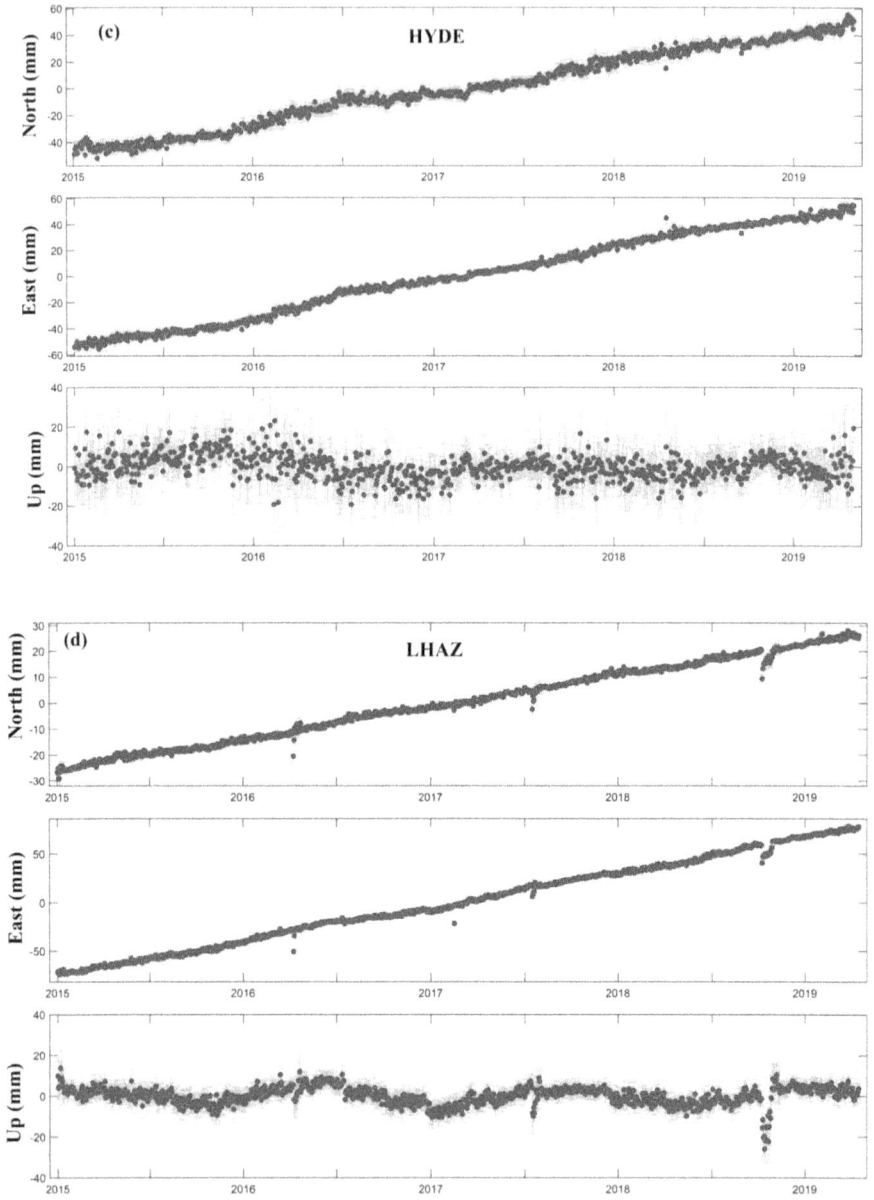

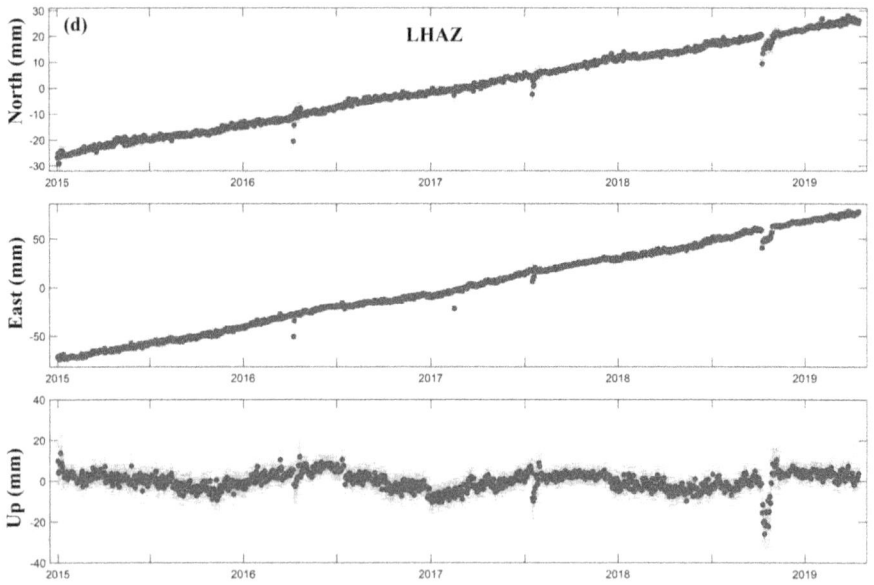

FIGURE 15.1   (Continued)

TABLE 15.1

GPS Velocities (in mm/yr) of All Four Stations Along with the IGS Reference Stations in the ITRF08. V_e, V_n, and U Represent the Station Velocity in East, North, and Upward Direction Along with Their Respective Uncertainties (σ_e, σ_n, and σ_u)

Site Name	Longitude (°)	Latitude (°)	V_e (mm/yr)	σ_e (mm/yr)	V_n (mm/yr)	σ_n (mm/yr)	U (mm/yr)	σ_u (mm/yr)
LCK4	80.9556	26.9121	36.12	0.24	34.81	0.06	−4.36	0.24
HYDE	78.5509	17.4173	39.97	0.07	34.68	0.06	−1.05	0.25
IISC	77.5709	13.0212	42.94	0.07	34.38	0.05	−2.53	0.24
LHAZ	91.104	29.6573	45.27	0.05	15.47	0.05	−0.76	0.21
CHUM	74.7511	42.9985	27.42	0.06	2.47	0.06	−2.23	0.21
KIT3	66.8855	39.1348	27.56	0.06	4.41	0.06	−1.21	0.24
POL2	74.6943	42.6798	27.21	0.04	4.61	0.04	−2.23	0.15
URUM	87.6007	43.8079	29.6	0.07	1.81	0.08	1.35	0.26

15.5 RESULTS AND DISCUSSION

The GPS processing results show that all GPS sites move with a velocity of about ~ 16–42 mm/yr in the east direction and ~ 35–45 mm/yr in the north direction (Figure 15.2 and Table 15.1). We have calculated the arc-normal velocities (i.e., perpendicular to the Himalayan arc) of all the four stations to evaluate the internal and plate boundary deformation of the Indian plate (Figure 15.3). We have observed that HYDE station moves northward with a rate of ~ 1.5 mm/yr relative to the IISC site (Figure 15.3). Similarly, LCK4 station converges towards the Eurasian plate with a velocity of ~ 1.2 mm/yr relative to the HYDE site (Figure 15.3). These relative surface velocities provide ~ 1 – 3 mm/yr horizontal deformation rate of India (south to the Himalaya). This result is consistent to the previous studies [e.g., Banerjee et al., 2008; Jade et al., 2004, 2014; Paul et al., 2001]. In addition, we have also observed that whole peninsular India is subsiding with a rate of ~ 3 ± 1 mm/yr. To derive the interplate deformation of the Indian plate, we have estimated the horizontal velocity (~ 48 ± 1.5 mm/yr) of LHAZ station. Comparing the arc-normal velocity of LHAZ with the other three stations (IISC, HYDE and LCK4), we have found that the internal deformation rate between India and Tibet turns out to be about ~ 15 ± 1 mm/yr. This deformation is mostly (about 50% of total deformation) concentrated along the Himalayan arc (north to the MHT) [Banerjee et al., 2008; Jade et al., 2014; Paul et al., 2001].

Low deformation rates (~ 1–3 mm/yr) along the plate interior show rigidness of the Indian subcontinent (Figure 15.3). These low rates insinuate the rare occurrence of earthquakes in the stable Indian plate. However, the central and southern parts of India have been struck by a few hazardous events in the past, implying the steady deformation of the Indian plate. In connection to this unusual phenomenon, Banerjee et al. (2008) suggested that the motion of the Indian plate could be separated into the

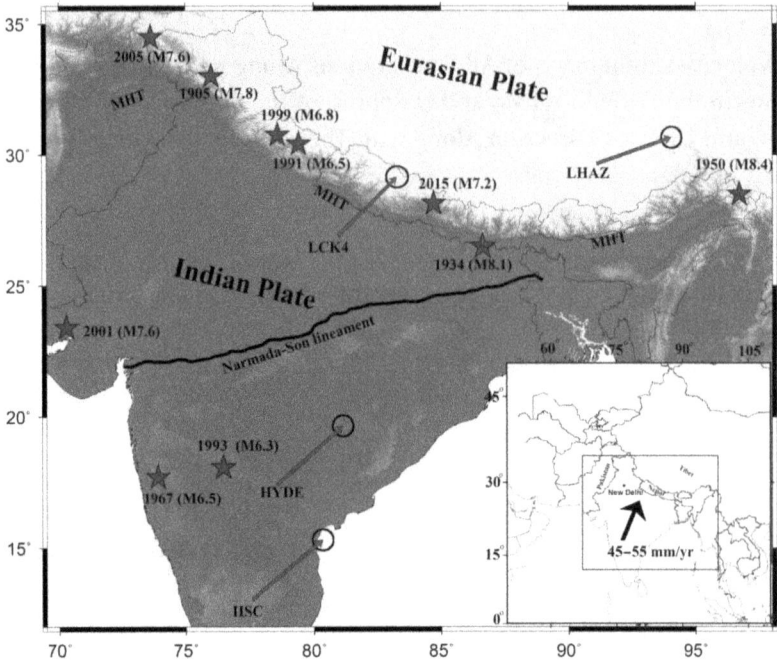

FIGURE 15.2   Surface velocity filed along the Indian subcontinent in the ITRF08. Blue arrows indicate the horizontal velocity of the four GPS stations. Small black circles are the 95% confidence error ellipses of GPS velocities. The blue star represents past large earthquakes with their respective magnitude (in bracket) along the plate interior as well as along the Himalaya. The solid black line indicates the Narmada-Son lineament. The red line represents the Main Himalayan Thrust (MHT). The black rectangle in the inset figure indicates the boundary of the main figure.

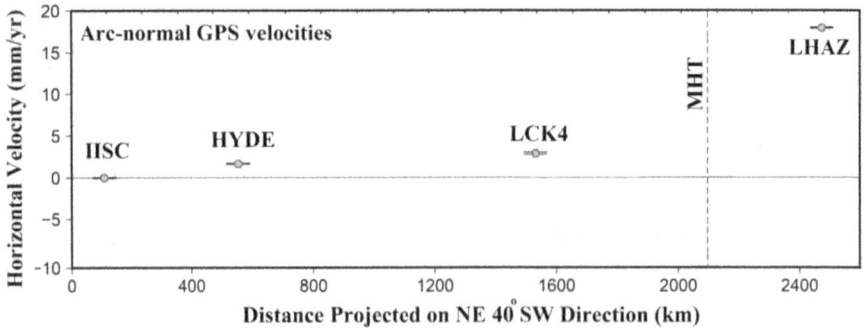

FIGURE 15.3   Arc-normal velocity profile of the studied GPS stations. The velocity difference of all stations in the normal direction has evaluated by fixing the IISC station situated on the stable Indian plate. All the velocities are projected in the NE 40° SW profile. The dotted black line indicates the MHT.

motion of two plates: southern Indian plate and northern Indian plate detached by the Narmada-Son lineament. In this setting, although the model fits the observations data, the relative motion of two plates shows a statistically insignificant change in the surface velocities [Banerjee et al., 2008]. However, Mahesh et al. (2012) tested the same hypothesis based on their GPS measurements and reported that the Indian plate could not be segmented into two or more plates [Mahesh et al. 2012]. Hence, due to the insignificant internal deformation of the stable Indian plate, the present study suggests that tectonic stress could be the main cause of the frictional failure of the plate interior [Zoback et al., 2002]. This means that, the occurrence of past intraplate earthquakes within India may be considered as produced either by the perturbations of the stress of lithospheric in the interior plate or by the compressive plate boundary stress from the Himalayan arc [Banerjee et al., 2008; Sharma et al., 2018, 2020].

15.6   SUMMARY

We accrue GPS data from four continuous IGS stations; three of them are established along the Indian plate (HYDE, LCK4 and IISC), and one is installed on the Eurasian plate (LHAZ). We utilize GAMIT/GLOBK post-processing software to analyze these GPS data. We obtain estimated velocities of these four stations in the north, east and the upward direction. The horizontal velocities of the Indian-plate stations lie between ~50 and 55 mm/yr in the northeast direction, whereas the vertical velocities of these sites lie between ~ −1.05 and −4.36 mm/yr. In addition, the surface velocity of LHAZ station (~48 ± 1.5 mm/yr) shows oblique motion towards east direction along with minor subsidence rate (~ −0.76 mm/yr). Using these velocities, we interpret the minor internal deformation of the Indian plate (~1–3 mm/yr) as well as the large deformation along the plate boundary (~ 15 ± 1 mm/yr). The increased deformation rates along the plate boundary suggest higher seismic hazard along the Himalayas. In contrast to that, the lower displacement rates along the plate interior support the rigidness hypothesis of the Indian plate [Banerjee et al., 2008; Mahesh et al., 2012]. However, due to lithospheric stress or stress generated from the Himalayas, the possibilities of a large earthquake in the future along the plate interior are undeniable [Zoback et al., 2002]. As a future work, re-analysis based on the dense GPS coverage along the Indian subcontinent could provide more constraints on the heterogeneity of the crustal deformation and associated seismic hazard estimation in the study region.

REFERENCES

Altamimi, Z., L. Métivier, and X. Collilieux. 2012. ITRF2008 plate motion model. *Journal of Geophysical Research: Solid Earth*. 117: B7.

Ambraseys, N., and J. Douglas. 2004. Magnitude calibration of north Indian earthquakes. *Geophysical Journal International*. 159: 165–206.

Avouac, J.-P., L. Meng, S. Wei, T. Wang, and J.-P. Ampuero. 2015. Lower edge of locked Main Himalayan Thrust unzipped by the 2015 Gorkha earthquake. *Nature Geoscience*. 8: 708.

Banerjee, P., R. Bürgmann, B. Nagarajan, and E. Apel. 2008. Intraplate deformation of the Indian subcontinent. *Geophysical Research Letters*. 35: 18.

Bilham, R. 2019. *Himalayan Earthquakes: A Review of Historical Seismicity and Early 21st-Century Slip Potential*. Geological Society, London, Special Publications. 483: 483–16.

Bilham, R., R. Bendick, and K. Wallace. 2003. Flexure of the Indian plate and intraplate earthquakes. *Journal of Earth System Science*. 112: 315–329.

Bilham, R., F. Blume, R. Bendick, and V. K. Gaur. 1998. Geodetic constraints on the translation and deformation of India: Implications for future great Himalayan earthquakes *Current Science*. 74: 213–229.

Bilham R., V.K. Gaur, and P. Molnar. 2001. Himalayan seismic hazard. *Science*. 293: 1442–1444.

Bilham, R., K. Larson, and J. Freymueller. 1997. GPS measurements of present-day convergence across the Nepal Himalaya. *Nature*. 386:61–64.

Blewitt, R., D. Lavallee. 2002. Effect of annual signals on geodetic velocity. *Journal of Geophysical Research: Solid Earth*. 107: B11.

Dam, T. Van, J. Wahr, P.C.D. Milly, A.B. Shmakin, G. Blewitt, D. Lavallee, and K.M. Larson. 2001. Crustal displacement due to continental water loading. *Geophysical Research Letters*. 28: 651–654.

French, G.-T. 1996. Understanding the GPS: An Introduction to the Global Positioning System. Bethesda, MD: GeoResearch, Inc.

Grewal, M.S., L.R. Weill, and A.P. Andrews. 2007. *Global Positioning Systems, Inertial Navigation, and Integration*. Hoboken, NJ: John Wiley & Sons.

Gupta, H.K. 1993. The deadly Latur earthquake. *Science*. 262: 1666–1668.

Herring, T., R. King, and S. McClusky. 2010. Introduction to GAMIT/GLOBK, Release 10.4. Massachusetts Institute of Technology. http://geoweb.mit.edu/gg/

Hofmann-Wellenhof, B., H. Lichtenegger, and J. Collins. 2012. *Global Positioning System: Theory and Practice*. Berlin, Springer Science & Business Media.

Jade, S., B. Bhatt, Z. Yang, R. Bendick, V. Gaur, P. Molnar, M. Anand, and D. Kumar. 2004. GPS measurements from the Ladakh Himalaya, India: Preliminary tests of plate-like or continuous deformation in Tibet. *Geological Society of America Bulletin*. 116: 1385–1391.

Jade, S., T. Shrungeshwara, K. Kumar, P. Choudhury, R.K. Dumka, and H. Bhu. 2017. India plate angular velocity and contemporary deformation rates from continuous GPS measurements from 1996 to 2015. *Scientific Reports*. 7: 11439.

Kaneda H., T. Nakata, H. Tsutsumi, H. Kondo, N. Sugito, Y. Awata, S.S. Akhtar, A. Majid, W. Khattak, A.A. Awan, R.S. Yeats. 2008. Surface rupture of the 2005 Kashmir, Pakistan, earthquake and its active tectonic implications. *Bulletin of the Seismological Society of America*. 98: 521–557.

Lavé, J., and J.-P. Avouac. 2000. Active folding of fluvial terraces across the Siwaliks Hills, Himalayas of central Nepal. *Journal of Geophysical Research: Solid Earth*. 105: 5735–5770.

Leberl, F. 1978. *Observations and Least Squares*. New York, NY: Elsevier.

Mahesh, P., J. Catherine, V. Gahalaut, B. Kundu, A. Ambikapathy, A. Bansal, L. Premkishore, M. Narsaiah, S. Ghavri, and R. Chadha. 2012. Rigid Indian plate: Constraints from GPS measurements. *Gondwana Research*. 22: 1068–1072.

Paul, J., R. Bürgmann, V. Gaur, R. Bilham, K. Larson, M. Ananda, S. Jade, M. Mukal, T. Anupama, and G. Satyal. 2001. The motion and active deformation of India. *Geophysical Research Letters*. 28: 647–650.

Sharma, Y., S. Pasari, O. Dikshit, and K.-E. Ching. 2018. GPS-based monitoring of crustal deformation in Garhwal-Kumaun Himalaya. International Archives of the Photogrammetry, Remote Sensing and Spatial Information Sciences 42: 451–454.

Sharma, Y., S. Pasari, K.-E. Ching, O. Dikshit, T. Kato, J.-N. Malik, C.-P. Chang, and J.-Y. Yen. 2020. Spatial distribution of earthquake potential along the Himalayan arc. *Tectonophysics*. 791: 228556.

Van Diggelen, F. 2009. *A-GPS: Assisted GPD, GNSS, and SBAS*. New York, NY: Artech House.

Yin A. 2006. Cenozoic tectonic evolution of the Himalayan orogen as constrained by along-strike variation of structural geometry, exhumation history, and foreland sedimentation. *Earth-Science Reviews*. 76: 1–131.

Zoback, M. D., J. Townend, and B. Grollimund. 2002. Steady-state failure equilibrium and deformation of intraplate lithosphere. *International Geology Review*. 44: 383–401.

Index

For Product Safety Concerns and Information please contact our EU
representative GPSR@taylorandfrancis.com
Taylor & Francis Verlag GmbH, Kaufingerstraße 24, 80331 München, Germany

9 780367 517441